This book provides a clear and concise introduction to the
study of plants, the science known as Botany. After discussing briefly the different kinds of plants that exist today,
it looks in some detail at the structure and behaviour of
flowering plants and the special characters of the non-
flowering types. There then follows a classification of all
the plant groups, and final chapters touch on such important
aspects of the science as ecology and the effects that man's
activities have had on the vegetation of the world. A
bibliography is included for those who wish to pursue the
subject further.

TEACH YOURSELF BOOKS

Dr Elliott . . . has most successfully carried out his stated intention of outlining the scope of Botany.

The Naturalist

BOTANY

John H. Elliott
B.Sc., Ph.D., M.I.Biol.
Principal Lecturer in Biology, Leeds Polytechnic

TEACH YOURSELF BOOKS
Hodder and Stoughton

First printed 1958
Sixth impression (with corrections) 1973
Ninth impression 1976
Tenth impression 1978

Copyright © 1958
Hodder and Stoughton Ltd

ISBN 0 340 05528 6

Printed in Great Britain
for Hodder & Stoughton Paperbacks,
a division of Hodder & Stoughton, Ltd.,
Mill Road, Dunton Green, Sevenoaks, Kent
by Richard Clay (The Chaucer Press), Ltd., Bungay, Suffolk

CONTENTS

ACKNOWLEDGEMENTS

I would like to express my very sincere thanks to Mr Dennis Lloyd, now of the Lancastrian Secondary Girls' School, Chichester, for his generous assistance with the illustrations; to Miss F. M. Morris, B.Sc., M.I.Biol., of the Leeds College of Technology for her help and criticism during the preparation and finally to my wife for her unfailing patience and encouragement.

J. H. E.

The front cover illustration is taken from Figure 178, 'Stages in pollination of sunflower', in *New Biology for Tropical Schools* by R. H. Stone and A. B. Cozens, published by Longman's, Green and Company.

INTRODUCTION

The study of plants must always have been an important part of the life of man even when it was not carried out systematically. Much of it may only have been pursued so far as it concerned his welfare, so that the study was empirical. From the earliest times man has derived food, shelter (including clothing), medicine and decoration from plants. In these respects therefore it has been to his advantage to investigate the Plant Kingdom, and it would be true to say that most of his knowledge has come from the study of " useful " plants. But as man became more learned his general desire for knowledge increased and plants (and of course animals) were studied for the sake of knowledge as well as for economic improvement.

Thus in most of the ancient civilisations one can find some references to plants both by the written word and by drawings, and though a good deal of the information which has come down to us is imaginative rather than factual, it is in these references that we can find the beginnings of what we call Botany. Much of this earlier work was descriptive and classificatory, possibly as a guide to medical aids, and though some experimental work may have been done, it was only in much later times that systematic experimental work and organised observation led to a real knowledge of how plants live.

In this book an attempt will be made to give a brief outline of the scope of the science known as Botany, and it will be emphasised that it is a subject in which a great deal of direct observation and practical work can be carried out by anyone who is prepared to take a country or even suburban walk. At the end of the book will be found a list of other works, in many of which can be found instructions for practical methods, but the reader will be able to do a great deal with a hand lens (10 × magnification), a pair of mounted needles, a scalpel or razor blade and a pair of forceps. It is, of course, desirable to keep a record of what one sees, and this is even more useful if accompanied by drawings. If the subject is being studied with a view to examinations, these records become essential, but in any case they lend much more point to the observer's activities. Since, however, the mere prospect of drawing seems for many people to be an awesome one, the author does not propose to labour it to the extent of deterring the would-be student from enjoying the study of plants.

In studying plants it must be borne constantly in mind that they are living organisms, and that though we may provide general descriptions and fairly general rules of activity, the student must always be prepared for individual differences and peculiarities and must not discard all his ideas just because some cases seem to be anomalous. On the other hand, the idea of living organisms must not be carried to the degree of purposive activity which is found in man. It is wrong to say that a particular organ has been developed to carry out a specific function. We can only say that as a result of certain developments a particular organ has become adapted and suitable for the discharge of this activity. Too little is still known about the influences and mechanisms which have led to such developments and very often one finds structural modifications which do not seem to be associated with any activity of the plant. This approach, which is known as teleology, must be avoided by the serious student.

In the present work it is not possible to pursue any aspect in great detail whilst giving a general picture of plants. A list of books suitable for further reading is given at the end, and doubtless these will open further roads for anyone who finds himself interested in the subject.

FOREWORD TO 1973 IMPRESSION

In the present edition a few minor alterations have been made in the terminology, together with a modernisation of one or two physiological conceptions more consistent with current theories.

1

KINDS OF PLANTS

IF we look at the plants which grow around us in our gardens, in the fields and along the hedgerows, we shall probably gain one major impression—namely, that the greater number of them bear flowers. Some of these flowers, such as those of the Grasses, are not very obvious individually, but are conspicuous when clustered together. This impression of the widespread dominance of Flowering Plants would be gained in most parts of the world, though in some of the great northern forests there would be a preponderance of trees with needle-leaves and with " cones ".

It is only when we examine the plant population more closely that we find a range of plants other than Flowering Plants. Some of them are very simple, whilst others approach the Flowering Plants in complexity of structure, but none of them reproduces by what we call " flowers". At the present day therefore the Flowering Plants are dominant, and we regard them as the most advanced form of plant life.

It was not always so, and in the past other groups have occupied much more important positions than they do now. Perhaps the most familiar example of this is the dominance of Fern-like plants (though very often differing markedly from our present-day Ferns) during the ages which contributed to the Coal Measures. It is generally accepted that there has been a gradual evolution of highly specialised types among plants, just as there has been in animals, probably with a more continuous thread of structural similarity running through the plant groups than is the case with animals. Nevertheless we shall find a wide diversity of structure between the simplest plants and the Flowering Plants. The sequence is not a simple one, and it will be found that there are digressions which are difficult to fit into the developmental scheme. Detailed consideration of many of them is beyond the scope of this book, but examples will be quoted.

Thus we see that the plant life around us belongs to various stages of plant evolution, and the main existing groups are given below, the simplest types first.

A. Thallophyta

1. Algae. These include unicellular, filamentous and thalloid forms which are mainly aquatic in both fresh and salt water. All possess chlorophyll, though in some cases it is masked by other pigments.

2. Fungi. Mushrooms, toadstools, moulds, mildews, etc. None of them possesses chlorophyll, and the vegetative structure is at the most filamentous.

In this section we should include the Lichens (which are associations of Fungi and Algae) and the Bacteria (which are of very simple structure and the classificatory position of which is difficult to determine exactly).

B. Bryophyta

3. Hepaticae. The Liverworts.
4. Musci. The Mosses.

C. Pteridophyta

5. Filicales. The Ferns.
6. Equisetales. The Horsetails.
7. Lycopodiales. The Clubmosses.

The above groups together constitute the Cryptogams or seedless plants.

D. Spermatophyta

8. Gymnosperms.

   (i) Conifers: the Pines, Spruces, Firs, etc.
   (ii) Cycads: tropical palm-like types.

9. Angiosperms. The Flowering Plants.

   (i) Monocotyledons: Grasses, Bluebells, Daffodils, Palm-trees, etc.
   (ii) Dicotyledons: Dandelions, Buttercups, Apples, Oaks, etc.

Though almost every plant existing today belongs to one or the other of these groups (a few small groups have not been mentioned), investigation of the fossil record shows that other forms now completely extinct have flourished in earlier days. Many of these represented links between existing groups, or perhaps one should say between the then existing representatives of these groups. In many cases the present representation is much smaller than in these earlier periods and only the Flowering Plants are more numerous in species than at any other time. As an example of the change we may quote the Gymnosperm order Ginkgoales. From 30 to 180 million years ago this was a very large order with many species, but today the Ginkgoales are represented by a single species confined (in natural conditions) to certain parts of China.

Many factors influenced the rise and fall of various plant groups. Climatic changes, geological upheavals, etc., have all played their part, and much remains to be learned concerning the distribution of the various forms.

It must be realised that the development of new types is usually very slow, measured by our own life-span, and though change is still going on, it is too slow for us to appreciate the progress.

Because the Flowering Plant is the most familiar to most people, it is proposed to study its structure and behaviour first. Some features will be found to be common to the other groups, but a later chapter will be devoted to a brief consideration of the special characters of the non-flowering types.

2

THE FLOWERING PLANT

1. MORPHOLOGY—THE PLANT FORM

Obviously the first feature about a plant which strikes the observer is its external appearance. The major characters which we use in identifying and classifying plants are those of form—we know a cabbage by the shape and arrangement of its various parts, a bluebell by the colour, shape and position of its flowers, though subsequent familiarity may permit us to use more subtle means of identification. So it is of great importance that the botanist should be familiar with the general structure of the plant and with the diverse modifications which appear in association with differing functions and environments.

How shall we choose a " model " on which to base our descriptions? Almost every plant species could be said to have some peculiarity of its own (otherwise we could not separate it from its neighbours), but since there is so much in common, we can justifiably pick a familiar plant to illustrate the main features of most of them. It may be pointed out that there seem to be two fairly well-defined groups of plants around us: one consists of relatively small rather tender types, and the other of definitely woody types, often reaching great size. Further examination will, however, show that the basis of construction is the same in both woody and herbaceous plants.

Fig. 1 is a drawing of a common garden plant—the Wall-flower. Part of the plant is permanently below the ground and is not normally seen. This is the root, and more will be said about it later. The aerial part is called the shoot, and consists of the stem, which is continuous with the root, and the leaves, which are the typically green parts of the plant. At certain times the shoot will also show flowers, and later fruits and seeds.

Some of the principal characteristics of these parts can now be considered in greater detail.

(i) The Stem

The stem is the main organ of support in the plant, and the way it grows determines to a large extent the form of the plant.

In the greater number of cases the stem is a rigid structure holding the leaves and flowers up into the air. But in some cases the stem is weak and trails along the ground or climbs by means of supports; in other cases it is very short and bulky, and in still other instances the main stem of the plant is wholly below ground. These modifications are associated with certain types of growth behaviour which will be discussed shortly.

Certain features are, however, always apparent, and it will be seen that the apex of the stem—that is, the younger end—always **tends** to grow towards the light and usually away from the ground, and the effect of this will be seen when the physiology of the plant is discussed. Where the stem is underground there is usually an aerial flowering shoot.

When young, the stem is usually fairly soft, but as it grows older it becomes harder and thicker, thus gaining that rigidity which is necessary to support an increasing weight of leaves and flowers and ultimately fruits. In herbaceous plants the stem may never become truly woody, although there is no hard-and-fast demarcation.

All stems bear leaves of one kind or another, and in the Flowering Plant there are small structures called buds always

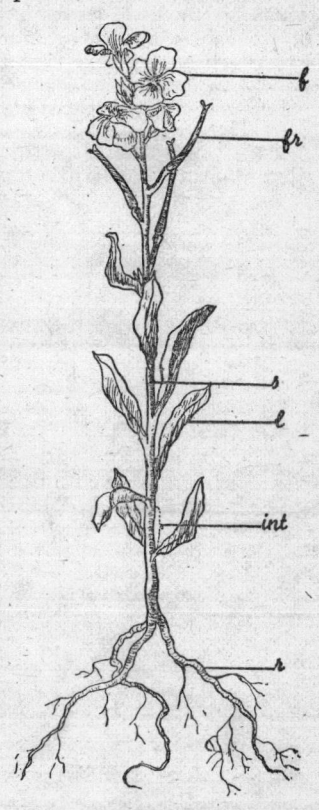

FIG. 1.—A Wallflower Plant.

f. flower, *fr.* fruit, *int.* internode, *l.* leaf, *s.* stem, *r.* root.

associated with the leaves. These buds are in fact miniature shoots, and they will elongate eventually into new stems with their leaves or flowers. In winter the plant may have died down to a small food-storage organ supporting a few buds (as in the potato), whilst on an old tree the leaves have gone and the buds represent the starting point for further growth in the spring. Most exposed buds are protected by special scales which are modified leaves or parts of leaves and which drop off when growth starts again in the spring, leaving a characteristic ring of scars, which thus marks the beginning of the year's growth on the woody twig.

Fig. 2 shows young twigs of Beech and Sycamore, and in

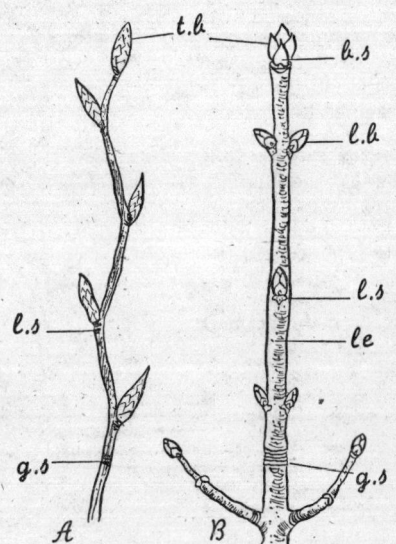

FIG. 2.—A. Twig of Beech.   B. Twig of Sycamore, both showing Monopodial Branching.

bs. budscale, *g.s.* girdle scar, *le.* lenticel, *l.s.* leaf scar, *l.b.* lateral bud, *t.b.* terminal bud.

Sycamore a single bud terminates the twig, whilst the lower buds are arranged in opposite pairs, as of course were the leaves with which they were associated. When that terminal bud starts to grow there will be a simple elongation of the existing twig, whilst the lateral buds may give shorter lateral branches.

Such growth is called monopodial or indefinite, and is typical of young plants. In some cases it may be the main type of growth throughout the life of the plant, even in an old tree. In the earlier stages at any rate it tends to give a tall, slim outline.

When for various reasons the terminal bud does not maintain the growth, the lateral buds become the new leaders. The commonest reason for the cessation of growth of the terminal bud is that it produces a flower, so that growth ends with the

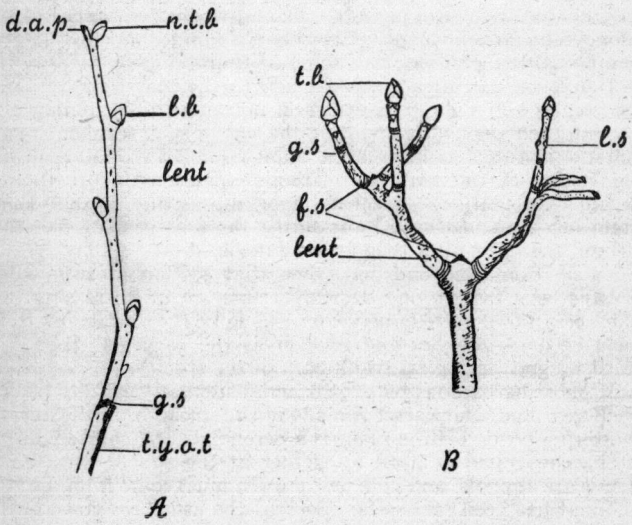

FIG. 3.

A. Sympodial Twig of Lime.

d.a.p. dead apical portion, *g.s.* girdle scar, *l.b.* lateral bud, *lent.* lenticel, *n.t.b.* new terminal bud, *t.y.o.t.* two-year-old twig.

B. Dichasial Twig of Sycamore.

f.s. flower-scar, *g.s.* girdle-scar, *lent.* lenticel, *l.s.* leaf-scar, *t.b.* terminal bud.

death of the flower. If the lateral buds are arranged singly, then further growth will be unidirectional, but (at any rate at first) at an angle to the original line, as shown in Fig. 3A, which is a young Lime twig. If the buds are arranged in pairs, then the simultaneous growth of the most distal pair will give the arrangement shown in Fig. 3B, an old Sycamore twig.

Both these forms of growth are called sympodial, and the double form is known as dichasial.

It is fairly safe to say that the general form of all Flowering Plants at least is determined by these methods of growth. In herbaceous plants the branching plan may not be obvious, often because of the early intervention of flowering, which imposes its own system (though this is very similar), but it may be followed with ease in woody plants, where persistence of a monopodial arrangement tends to give a slim form, whilst later sympodial branching gives a rounded form.

Reference has already been made to the scars left by the bud-scales and their significance in marking the beginning of growth each year. By counting the sets of such scars one can tell the age of a twig or even branch for a long time, until thickening of the bark and lateral expansion distort them beyond recognition. A most important feature is that the stem increases in length only during the first year of growth behind the apex—if a twig grows 2 inches during its first year of elongation from the bud, then that section of twig will always be 2 inches long, though it may go on increasing in thickness for hundreds of years. Usually the surface of the twig becomes rougher and rougher as the bark gets thicker, until we get the typical trunk-bark of the species.

In most herbaceous plants the aerial stem at least dies back each year and a new start is made in the spring from the over-wintering portion if the plant is a perennial. Of course many herbaceous plants die off altogether at the end of the year. These are annuals, and here new growth must come from seeds.

What has been said so far concerns the usual appearance of the stem. Further study shows that the stem is subject to a variety of modifications involving both appearance and function. These may be associated with food storage, vegetative propagation and climbing, and will be dealt with later.

(ii) The Leaf

Typically the leaf is a flattened structure having a large surface in proportion to its thickness. It will become apparent that leaf-form is closely associated with function and environment and that under special conditions the leaf will show anomalous forms. Figs. 4 and 5 illustrate some of the commoner leaf-forms. The simple leaf (Fig. 4) has a single flattened portion which is the lamina or blade. This is attached to the stem by a part known as the leaf-base, and, unlike the stem, this is the region in which growth continues longest and the leaf does not show apical growth. Between leaf-base and lamina there may be a stalk or petiole. If this is absent, the leaf is said to be sessile. The lamina may be of various shapes (spear-shaped or lanceolate, heart-shaped or cordate etc.), and the edge or margin may be smooth (entire) or divided in some way—e.g. toothed or wavy. A few examples of simple leaf-forms are shown.

The leaf is supported by strands of tissue in which the

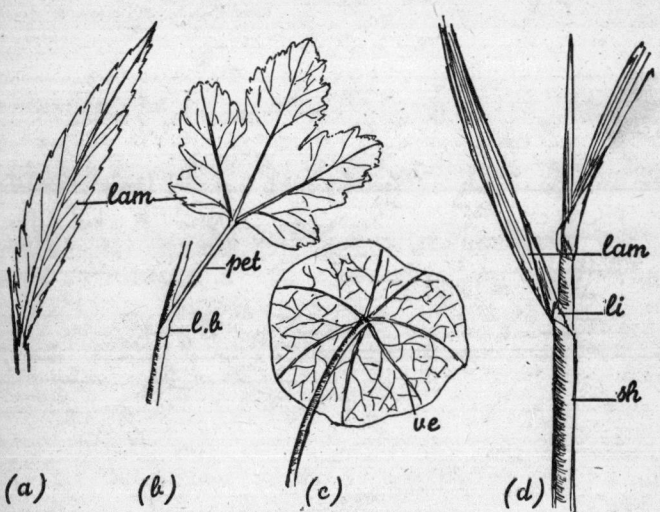

FIG. 4.—Types of Simple Leaves.

(a) Golden Rod, (b) Flowering Currant, (c) Nasturtium, (d) a Grass.

lam. lamina, *l.b.* leaf-base, *li.* ligule, *pet.* petiole, *sh.* sheath, *ve.* vein.

internal conducting elements are found These structures are called veins and again show distinctive forms. Sometimes the leaf has a single main vein (unicostate) and sometimes several (multicostate). As the veins pass towards the margins of the leaf they become less obvious, but the conducting tissue is still present within the leaf.

The angle which the leaf-base makes with the stem is called the axil, and in the Flowering Plant is always occupied by at least one bud. It is important to note this condition because it helps us to recognise some of the less obvious leaf-structures

and also how to distinguish between a simple leaf and the
leaflet of a compound leaf. The compound leaves are illustrated in Fig. 5, and it can be seen that the whole blade is
now completely divided into segments which are called leaflets
and which arise either from the top of the petiole or along a
continuation of the petiole. No bud is found at the base of
the leaflet. Two main types of compound leaf are found, the
pinnate and the palmate, and examples of each are shown.

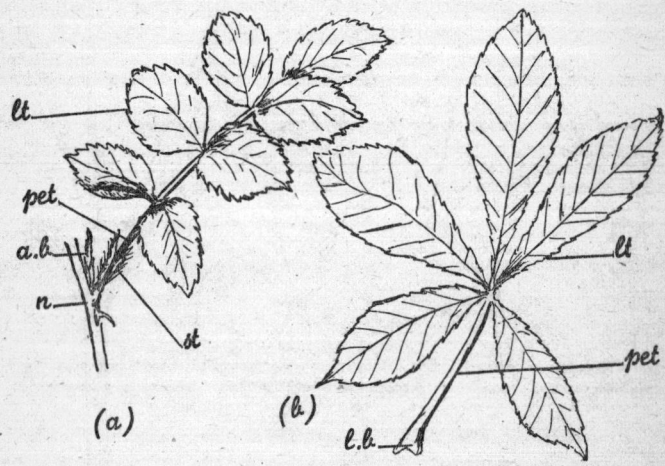

FIG. 5.—Types of Compound Leaves.

(a) Rose, pinnate.   (b) Horsechestnut, palmate.

a.b. axillary bud, l.b. leaf-base, lt. leaflet, n. node, pet. petiole, st.
stipule.

The structure of a leaflet is usually essentially similar to that
of a simple leaf.

All the leaves so far considered exhibit two distinct and
different surfaces and are called dorsiventral or bifacial. The
upper surface is technically the ventral surface and the lower
surface (the " back " of the leaf) the dorsal one. Either may be
hairy, but such a condition is more often associated with the
lower or dorsal side.

Other leaf-forms will be encountered, such as the elongated
leaf seen in Grasses and many other Monocotyledons, the

hollow cylindrical leaf of the Rush or the Onion and the much-divided leaves of submerged water-plants. Many of these leaves do not exhibit two different surfaces and those of the grass type are called isobilateral, a form also well illustrated by Daffodil, Bluebell, etc.

The leaves succeed one another on the stem in a sequence which is fairly constant in a particular species. This sequence is called the phyllotaxis. If several leaves arise at one point the condition is cyclic, and the most common type is the presence of opposite pairs, as in Deadnettle and Sycamore. When the leaves arise singly the phyllotaxis is said to be alternate, but the leaves are actually arranged in an ordered spiral manner. The rapidity of growth, and especially the number of leaves finally developed on a particular stem, may make it difficult to recognise a specific phyllotaxis.

So far reference has been made specifically to the green or foliage leaf, i.e. the structure which is the usual conception of a leaf. Other types of leaves, however, are commonly found. Thus underground stems frequently bear colourless scale leaves, and these may be well developed on storage organs such as bulbs. We have already seen the closely overlapping scales which appear as protective structures on the buds of trees, etc.

Leaf-like structures associated with flowering shoots are called bracts and will be dealt with later, and of course there is the great probability that the floral parts themselves are modified leaves.

All true leaves, however, have a bud in the axil.

Accessory structures are frequently associated with the leaves, and perhaps the most usual is the outgrowth of small organs called stipules near or from the leaf-base.

Like stems, leaves are subject to a good deal of modification, and some of these modifications will be considered shortly.

(iii) The Root

The most obvious function of the root is that it anchors the plant in the ground, and plants without root-like structures are usually aquatic, though even in this environment many of the floating plants have roots. Besides this function, however, the root is mainly responsible for the intake of substances into the plant, except in some of the lower forms.

The root starts as a slender colourless organ which characteristically grows into the ground, and particularly towards water. Like the stem, its growth is apical, but there is no bud at the tip nor are there any superficial structures such as leaves or lateral buds. The apex is protected by a covering known as the root-cap, which is constantly renewed as contact with the soil wears it away.

The region of elongation in a root is very short, compared with that of the stem. Shortly behind the tip and towards the upper limit of the growing region there is a region covered with fine outgrowths called root-hairs. These are most important because all absorption is carried out through them, and since they are destroyed as the root grows, it means that the plant must continually be producing new rootlets. Even in perennial plants, therefore, root-growth is likely to be much more continuous than stem-growth. Behind the root-hair region we first see the branch roots or lateral roots, and thus these are formed in a part of the root which has definitely stopped elongating. They recapitulate all the features of the main root. The older roots become thicker, and in a tree will become woody, thus resembling the branches, though they are probably never of the same dimensions. The area covered by the roots of a tree is often very extensive.

The form of root growth may be compared with that of the stem. Thus if the original root continues to dominate growth we get what is called a tap root, and this is very well seen in Dandelion and Dock and in an exaggerated form in Carrot and Beetroot. Lateral growth is then usually insignificant. On the other hand, if the main root dies back or fails to maintain its lead the lateral system expands and we get a mass of laterals forming a fibrous root system (Peas, Beans, Groundsel, etc.). In many plants new roots arise from stems, leaf-bases, etc., if these are in contact with the ground, and such roots are said to be adventitious. They too usually provide a fibrous system, and this type of development can be seen very well in Grasses. Root-systems may be shallow or deep, depending often on the availability of water, and they tend to remain active even when the aerial part of the plant has died down. Sometimes they produce special buds (adventitious buds), and thus can maintain growth of the plant. Roots do show some modifications, but on the whole are less variable in form than other plant organs—probably because the soil is a more stable environment. Figs. 6–8 show some typical root-forms.

(iv) Modifications of Stem, Leaf and Root

So far the general features of the main parts of the plant have been discussed. An examination of the plants around us shows that the normal form is often modified in association with function and environment.

In particular, modifications are to be found in connection

with food storage, perennation (persistence through the non-growing season), vegetative propagation (and often a combination of the three), with climbing and with a variety of ecological conditions. Not infrequently the modifications

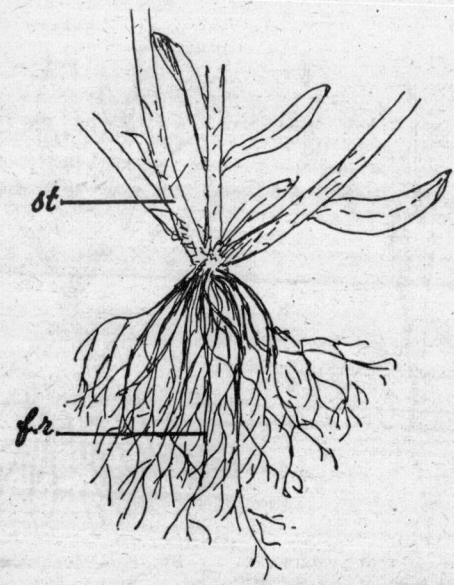

FIG. 6.—Fibrous Root System (Forget-me-not).
f.r. fibrous roots, *st.* stem.

involve more than one part of the plant, and may serve more than one purpose. In the following review of such organs examples will be given as far as possible to illustrate the various points discussed.

Vegetative reproduction or propagation

This process involves the development of new plants by an eventual separation of some part of the old plant. Almost invariably the process is initiated by the growth of a bud which is carried away from the parent by the elongation of a stem, or becomes surrounded by the organs of the new plant whilst still in close association with the parent. Separation is

achieved in one case by the withering of the connecting stem
and in the other by the breakdown of the old parent tissues,
and this may take place in a number of ways. In view of the
necessity for supporting the new individual, it will be realised

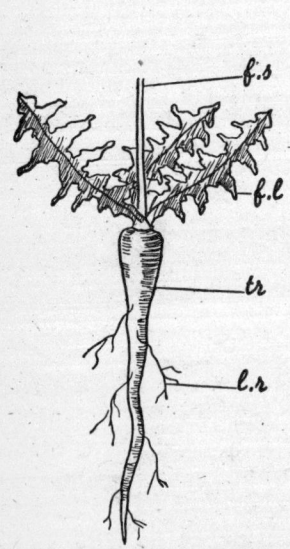

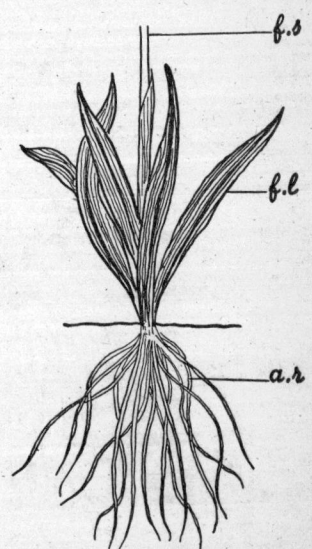

FIG. 7.—Tap-root System
(Dandelion).

f.l. foliage leaf, *f.s.* flower-
stalk, *l.r.* lateral root, *t.r.* tap-
root.

FIG. 8.—Adventitious Root
System (a Grass).

a.r. adventitious roots, *f.l.* foliage leaf, *f.s.* flowering shoot.

that such structures frequently involve the accumulation of
food deposits, and many of these have been developed for the
use of man.

Some of the commoner types are included in the following
descriptions.

The Strawberry (Fig. 9) propagates regularly by **runners**,
which are long stems arising from buds in the axils of leaves on
the parent plant, usually after flowering. They grow rapidly
away from the main plant and bear only a scale leaf and the
terminal bud. In contact with suitable soil the latter eventually sends out adventitious roots, and a new plant becomes

established. The process is then repeated from the daughter plants, so that a series of new individuals is quickly formed, and these are isolated by the withering of the runner. No special storage of food is involved.

There are many variations of this process. In the House-leek and the Daisy, for instance, very short runners called offsets are formed and soon produce a closely packed colony of plants, though there is rarely a second production from the daughter plants in the same season. The Blackberry scrambles over hedgerows, and in the autumn, during or after fruiting,

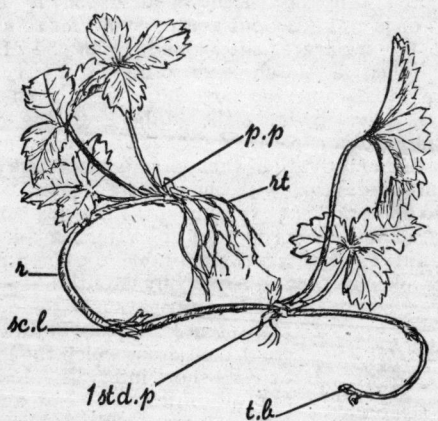

FIG. 9.—Strawberry, showing Runners.

1st. d.p. 1st daughter plant, *p.p.* parent plant, *r.* runner, *sc.l.* scale-leaf, *t.b.* terminal bud.

the long, straggling stems droop towards the ground, and when they touch the soil the terminal bud roots, growing up the following spring to form a new plant. Such a propagating structure is called a stolon, and the Blackberry can rapidly encroach from a hedge on to cultivated land.

Some plants have a permanently trailing habit, and the growth of lateral stems help in propagation, in that when they die or are broken the terminal part which is left becomes a new plant. Included in this type of growth are species such as Ground Ivy, Yellow Pimpernel and Creeping Cinquefoil.

In many other cases the stem is modified by the deposition of reserves of food material which may be associated with the formation of a number of buds. The way in which growth of these buds takes place leads to the formation of new plants in many cases, whilst in others the system provides a structure capable of withstanding adverse weather conditions (which in this country at any rate means winter), and thus provides for perennation. Needless to say, perennation and propagation are often closely associated.

One of the most familiar examples of this type of growth is the Potato. In this plant a special organ is formed, called a stem tuber. It arises from a lateral bud at ground level (or slightly above) on the parent plant. Usually a number develop at once and grow out as thin stems for some distance parallel to the surface of the ground. The end of the stem becomes swollen as starch is deposited in the cells of the developing tuber, and the connecting stem eventually dies away. The lateral buds on the "tuber" are the eyes of the potato, and each is capable of producing a new shoot, a fact which is used in cultivation when large potatoes are cut before planting into pieces each with one or more "eyes". Potatoes are earthed up to prevent exposure of the growing tubers to the light, as this causes them to turn green, and in this condition they are unfit, or at any rate unsuitable, for consumption.

In the case of the Crocus, Arum Lily, Montbretia and others, the organ of perennation and propagation is the corm. The main stem of the plant is always a short, stout body from which the leaves arise at very short intervals. Lateral buds appear during the year, and as new materials are synthesised by the foliage leaves they are transferred to the bases of these buds (but in particular the terminal one), giving rise to new corms. In addition, some of the food from the old corm may also be transferred to the new ones, whilst new shoots are preformed in the buds ready for early growth in the following spring. When these new shoots begin to grow the old corm has usually shrivelled away (except for instance in Montbretia, where a chain of corms is produced), so that the new plants are free. A Crocus corm is illustrated in Fig. 10.

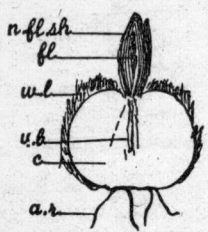

FIG. 10.—Crocus Corm. Vertical Section.

a.r. adventitious roots, *c.* corm, *fl.* flower, *n.fl.sh.* new aerial shoot, *v.b.* vascular bundles, *w.l.* withered leaves of last year.

Probably the most common stem modification is the **rhizome.** This is an underground stem which may remain slender or may be greatly enlarged by the storage of reserve materials. It is frequently the main stem of the plant, sending up an aerial flowering shoot annually, as is the case in Dog's Mercury and Solomon's Seal, or having a terminal group of leaves, as in Iris and many Grasses. The rhizome may show several years' growth, due to the fact that more food is accumulated during the year than is subsequently used. This is illustrated in Solomon's Seal, shown in Fig. 11. Such structures are mainly perennating organs, and propagation is usually casual, due to the accidental separation of a lateral branch.

So far the modifications described have been confined mainly

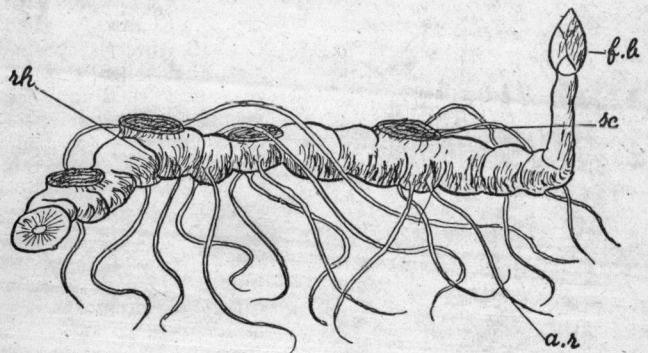

FIG. 11.—Rhizome of Solomon's Seal.

a.r. adventitious roots, *f.b.* flower-bud, *rh.* rhizome, *sc.* scar of old flower-shoot.

to the stem, but others also involve leaves or parts of leaves. Of these the most commonly occurring is the bulb. Bulbs are typically Monocotyledon structures, since their formation depends on the close arrangement of a number of leaves on a very short stem, a structural condition which is very frequent among the Monocotyledon herbs. The food reserves are now deposited in the leaves—in the case of Daffodil, Onion, etc., the bases of the actual foliage leaves being used, whilst in Tulip, various Lilies and others special scale leaves are the storage organs. Buds are of course present in the axils of such leaves, and some of these will develop to give the aerial shoots of the following year. As these shoots grow each begins to form its own bulb, whilst unused food from the parent bulb may be

transferred to the new ones. An Onion bulb is shown in
Fig. 12. On the outside of the bulb there are old withered
leaf-bases from which the food has gone, and these enclose the
fleshy bases, which are still full of reserve material (mainly
cane-sugar in the Onion). Two new shoots are shown, of
which the central one will form the new main bulb in the next
year of growth, whilst a lateral bulb will form around the
axillary shoot, though this bulb may not separate for two years.

Finally we may refer to such structures as **bulbils,** which
are usually buds associ-
ated with one or more
adventitious roots which
have become packed with
food reserves. In such
cases the bulbils become
separated when the old
plant dies down, and they
produce new and indepen-
dent plants the following
year. An example of this
type is shown in the illustration of the Lesser
Celandine in Fig. 13. On
the other hand, there are
bulbils which develop in
different circumstances,
as for instance in certain
species of Onion, where
the flower-buds develop
into bulbils and these
just drop off during the
autumn.

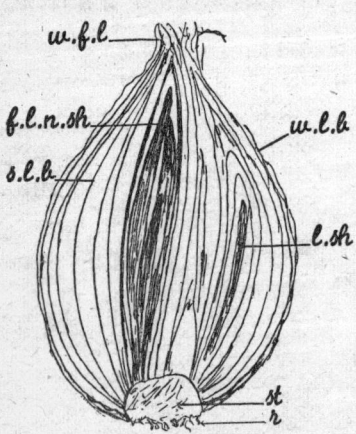

FIG. 12.—Vertical Section of Onion
Bulb.

f.l.n.sh. foliage leaves of new shoot,
l.sh. lateral shoot, *s.l.b.* swollen leaf-
base, *st.* stem, *r.* roots, *w.f.l.* withered
foliage leaf, *w.l.b.* withered leaf-base.

In all these modifica-
tions the importance of
the bud must be empha-
sised, and it will always be found that the special developments
for perennation and propagation are built round the bud,
which is the embryonic shoot and will provide the new plant.

Climbing modifications

An examination of most hedgerows will show a number of
plants which support themselves by clinging to more rigid
supports, particularly woody members of the hedgerow flora.
These climbing plants invariably have a stem which is incapable of erect growth and will sprawl over the ground until

they reach some means of pulling themselves up towards the light.

In general, this upward movement is achieved by special growth action in stems or leaves which results in the coiling of the organ around the support. This type of growth is generally associated with the stimulus of gravity in some cases and of contact in others. Gravity always produces some degree of spiral growth, but this is not noticeable in a rigid stem, though it becomes very obvious in a weak one. In the case of organs which respond to contact stimulus, little curvature takes place until such contact is provided, and it is interesting to note that it must be provided by a solid object, since the touch of raindrops, etc., seems to have no effect. In addition to climbing, which is achieved by specific growth activities, there are plants which simply scramble over the support (or obstacle) by means of superficial hooks or prickles, such examples being found in the Black-berry and Rose. In the case of Ivy the support is held by means of adventitious roots along the stem of the Ivy which penetrate into crevices.

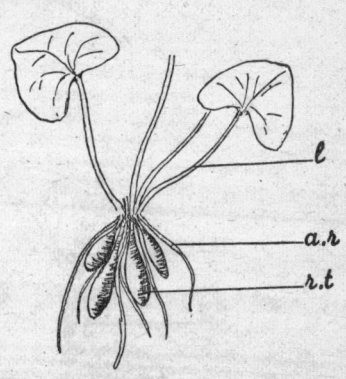

FIG. 13.—Bulbils of Lesser Celandine.

a.r. adventitious roots, *r.t.* root tuber, *l.* leaf.

Examples of climbing plants are readily found: the Scarlet Runner Bean (Fig. 14) in the garden, Convolvulus, Black Bryony and Wild Hop in the hedges, together with the less obvious Honeysuckle, which consolidates its position, so to speak, by developing a woody stem, are all examples of climbing by means of a twining stem. Here the stem apex grows along an exaggerated spiral path, and if this path brings it around a solid support, then the whole stem will be held firmly because once the curve of the stem has taken place it is permanent, and became more rigid as the stem gets older. The apex follows a clockwise direction of growth in some plants and an anti-clockwise path in others, but it is always the same in any one species. When a number of such stems become entwined, a thick rope of considerable rigidity is produced.

The other principal method of climbing is by tendrils, delicate and often quite beautiful structures exemplified in such plants

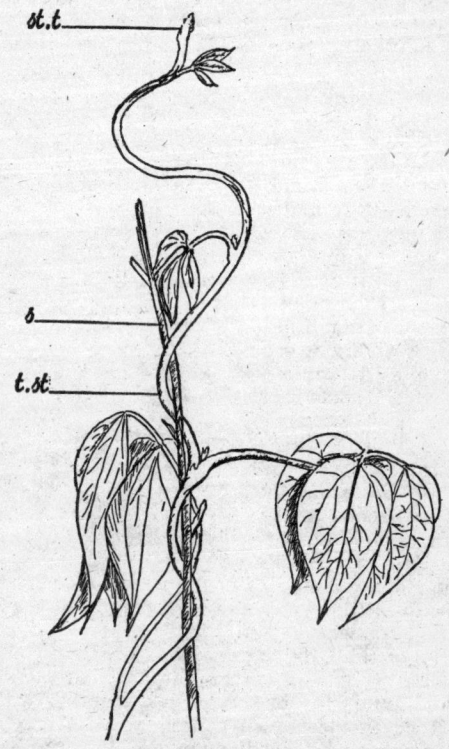

Fig. 14.—Twining Stem of Runner Bean.

s. support, *st.t.* stem tip, *t.st.* twining stem.

as the Sweet Pea (Fig. 15), Garden Pea (and many other members of the family), Clematis, Cucumber, Grape and many others. In the Cucumber, Grape, White Bryony and Passion Flower the tendrils are modified stems, or perhaps one should say more accurately, shoots, whilst in the Peas and in Clematis the tendrils are developed from leaves. In the tendril the young structure is hooked at the tip, and contact with a solid object on the inner surface of the hook will cause the latter to become more curved, probably because growth on the inner surface is inhibited. This curvature keeps the inner surface of the tip (which is the only sensitive portion) in contact with the support, so that a coil is gradually formed. Here again it is only the terminal part of the tendril which is actually

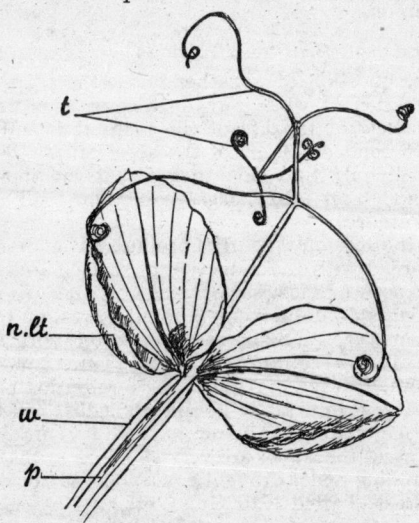

FIG. 15.—Leaflet Tendrils of Sweet Pea.

n.lt. normal leaflet, *p.* petiole, *t.* tendrils, *w.* wing.

elongating and curving, but the coil is fixed as the older tissues become more rigid. Sometimes the direction of growth reverses after a time so that the coil is formed in the opposite direction, and this not only gives further strength but also produces a springiness which tends to resist wind action and perhaps the weight of fruits, which might otherwise snap the stem. Continuous contact is not necessary, but the growth effect is not immediate, so that where a tip repeatedly brushes a support a spring-like coil often forms, although the support is not gripped.

A variant of this procedure is seen in Virginia Creeper, where the tips of short lateral shoots are flattened into discs, which when pressed against a stone surface or other support adhere tightly by suction and support the main stem. That the contact is very firm is shown by the fact that if the stem is pulled away, bits of the support may be brought with it.

Most methods of climbing likely to be encountered by the reader will come within the scope of those described, and it must be emphasised again that the twining stem and tendril represent definite growth movements started as a response to external stimuli.

Climatic and ecological modifications

It may be questionable whether this is the proper place for the consideration of what is really a very wide field, but in our brief review of the life of the plant it is unfortunately necessary to condense some of the aspects. At the moment reference will only be made to some of the morphological features, though in other pages something will be said of internal differences.

Perhaps the most frequently encountered modifications are those concerned with the reduction of surface area in connection with exposure to excessive water loss, etc. In many such cases the plants do not develop normal leaves, and it is found that the stems perform many of the functions normally carried out in the leaf. Examples of such plants are the Cacti, Gorse, Broom, growing in regions where water loss is very high or where water is difficult to obtain, though it will be obvious that these plants are often found in more favourable areas.

In other cases the leaves are very small, as in the Heaths, or capable of rolling under drought conditions, as in many grasses (Marram Grass, Tufted Hair Grass, etc.). Not only are such modifications found in dry places or where evaporation is high, but they are found in salt marshes where the high concentration of salt makes water absorption difficult. The modifications associated with all these conditions are referred to as **xerophytic**.

On the other hand, aquatic plants also show departures from the normal form. Thus totally submerged plants or parts of plants have very finely divided leaves, which are therefore not subject to pressure from water movement, and the floating leaves are large and often provided with mucilage glands.

Lastly there are the plants which have modified leaves for the catching of insects as a method of supplementing their nitrogen supply, and further reference will be made to these later.

In general it is true that the leaf is the part most frequently affected by these ecological modifications, or at any rate the structural changes are the most obvious. It must be realised, however, that not all peculiarities of structure can be interpreted in terms of special functions and that in many cases no apparent reason can be found for the oddities of structure which are to be encountered.

2. ANATOMY—THE INTERNAL STRUCTURE
(i) The living cell

Most people are to some extent familiar with the external appearance of plants but probably have little idea of how the plant is constructed.

The basis of all living organisms is protoplasm—a complex system of organic material the properties of which are still not fully known, although the development of cell and molecular biology has produced a great advance in information. The study of the living material is made difficult by the fact that any disturbance of its normal condition may alter its appearance and properties and the classical method of examining dead, stained tissues left much to be desired. This is also one aspect which has to be balanced against the advantage of the enormous magnification and detail revealed by the electron microscope. Phase contrast and interference microscopes have contributed markedly to the study of *living* cells.

Except in a few cases this living material does not occur in large masses but is divided into small units which in the plant are normally enclosed in a non-living cell wall consisting primarily of cellulose. These units are called cells, and plants range from single-celled individuals to large organisms containing millions of cells. Even the latter start as a single unit formed by the union of the male and female germ cells. This unit has the ability to reproduce its living material by replication, and this is distributed by repeated division into daughter cells, so rapidly increasing the cell mass. All these cells are derivatives of the original single unit—an important fact in the individuality of any plant.

Protoplasm consists mainly of proteins containing the chemical elements carbon, hydrogen, oxygen, nitrogen and, in many cases, sulphur and phosphorus. The structure is very complex and it should be pointed out that the term " protoplasm " is now rarely used in its original implication. Instead one commonly refers to " nucleus " and " cytoplasm " as the main regions of the cell. The cytoplasm consists of all the non-nuclear components, and the various specialised parts of the cytoplasm are referred to as organelles. Each of these has a distinctive structure and function. In general the cytoplasm seems to be more fluid and mobile in active cells and can be observed to flow. In resting and storage cells the cytoplasm appears much more jelly-like and these changes are consistent with a colloidal condition, so that we are dealing with a biological colloidal system.

The term "cell" probably originated with Robert Hooke, who, when experimenting with an early microscope, saw the box-like structure of cork, which is a plant material. The term was somewhat unfortunate from the modern viewpoint because what Hooke saw were the dead cell walls, whilst we apply the term cell to the living contents. In the 1830s Schleiden and Schwann propounded the theory that all living organisms were ultimately composed of these living units and so established the "Cell Theory". The term "typical cell" is often met but is difficult to define—the embryonic cell produces many successors which alter in type as they mature but nevertheless all retain certain characteristics.

Cytoplasm normally appears as a semi-transparent substance somewhat granular in consistency. In most specialised plant cells it does not fill the cells but encloses fluid-filled spaces called vacuoles. This fluid is the cell sap which permeates all the living material and contains various substances in solution or suspension. The cytoplasm is closely associated with the cell wall, which in fact it has produced, and in many living cells continuity is maintained from cell to cell by the existence of fine cytoplasmic threads which persist from the original cell division and are probably the site of the pits in xylem vessels.

Special surface membranes at the free boundary of the cytoplasm, of the nucleus and many of the organelles constitute an important aspect of cell structure. In addition, a much folded membrane of similar structure, continuous with the nuclear and cytoplasmic membranes, is called the endoplasmic reticulum. All these membranes appear to consist of an inner and outer layer of protein enclosing a double layer of lipid (fatty) molecules. This combination possesses very definite properties, including the most important one of regulating the entry and exit of dissolved substances, though various mechanisms may be involved.

Dominating the cell is a body called the nucleus. Visually it appears as a more or less spherical body more densely granular than the general cytoplasm. One of its properties is that it is easily stained by various dyes, and because of this the nuclear material is called chromatin. Normally this appears to form a network within a nuclear membrane, but at certain times, especially when the nucleus is going to divide, the chromatin appears as a number of separate bodies called chromosomes, of which there is a fixed number in the nucleus of every vegetative cell of a particular species.

It is now accepted that deoxyribonucleic acid, better known as DNA, is the vital constituent of chromosome structure and controls all cell activities and thus those of the organism. The control is transmitted via another nucleic acid — ribonucleic acid or RNA, which, unlike DNA, can pass freely into the cytoplasm and is vitally concerned in metabolism. Concentrations of RNA occur in the nucleoli.

Typically a cell contains only one nucleus, but in some of the lower plants binucleate or multinucleate " cells " are frequently observed. In such cases the term cell has probably not got the same significance as it has in the higher plants.

Of the living contents of the plant cell perhaps the most obvious are the plastids, which contain pigment and are typified by the chloroplasts. In addition, however, there are the finer structures known as mitochondria, ribosomes and the Golgi apparatus, all of which have important, though not yet fully understood, activities.

It may be pointed out that the cytoplasm is not at rest in the cell and in some cases and under some conditions active streaming movements are quite usual. Not only that, but the cytoplasm is capable of passing rapidly from the inactive, more rigid **gel** state to the more fluid **sol** condition.

Cells also contain many non-living substances, which are usually products or by-products of synthetic activity. Thus starch grains and oil globules are examples of substances synthesised for the use of the plant, whilst many mineral crystals are waste by-products.

It has already been said that the properties of the cell change as it grows older. Thus the juvenile cell is capable of division, but this property is lost as the cell reaches its adult form, though it may be regained in certain special circumstances. This change from the juvenile to the adult condition is called differentiation and leads to the appearance of the different cell masses or tissues seen in the older parts of the plant. Differentiation involves changes not only in cell shape and activity but also in the nature of the wall. Additional material is deposited and chemical changes take place, so that the staining reactions and the structure of the wall are quite different. In many cases the final stage of differentiation is the death of the cell, but this does not mean the end of its usefulness to the plant, and the wall changes may go on long after the cell is dead. To a considerable extent the differentiated form of the cell has a direct bearing on its functional activities.

In general, juvenile cells are restricted to certain parts of the plant which are called meristems, to which further reference will be made very shortly.

In the higher plants the apical meristems consist of a group of similar cells, but in many of the lower forms the whole cell system can be traced to a single apical cell.

At the apex the cells tend to divide in all planes, so that the organ increases in thickness as well as in length. Then there is a region of vigorous elongation due to the intake of much water accompanied by a preponderance of transverse divisions. Gradually vacuoles appear in the cells as the formation of protoplasm falls behind the expansion of the cells and new wall material is constantly being deposited to keep pace with increase in size. All this is differentiation, and the cells gradually assume their adult form. The conditions which govern differentiation are very complex and the various factors concerned may alter so that the development of an organ or the whole plant may be altered.

Cell division—the formation of new cells

Before going on to discuss the structure of the adult tissues, it is desirable that further details should be given of the processes by which new cells are produced. Emphasis on the constancy of the chromosome number in a species is most important in the light of the acceptance of the chromosomes as sources of the genes or hereditary factors.

In the ordinary vegetative cell of the plant there are two sets of similar chromosomes, derived in the first place from the male and female germ cells produced by the parents. This double set is referred to as the **diploid** or $2n$ number (i.e. there are n pairs) and is maintained throughout the whole body of the plant except in circumstances which will be described. At some time prior to the formation of germ cells (though, as we shall see later, this has to be interpreted very broadly) the chromosome number is reduced to a single set, which is the condition in the germ cell itself.

The formation of new cells is general throughout the young embryo and is initiated by the first division of the fertilised egg-cell. As the mass of cells so formed begins to differentiate, division becomes more restricted to the juvenile region and slows down in the cells which are becoming adult, finally ceasing altogether.

These juvenile regions are the meristematic areas already referred to, and in most multicellular plants are present at the apex of stem and root or of the thallus in plants not differentiated into stem, root and leaf. These apical meristems are directly descended from the dividing embryo and have always been juvenile. But other meristems which may be called lateral meristems develop in some regions from differentiated cells. This includes the cambium in dicotyledon roots and the cork cambium in general. Wounding will also stimulate new meristematic activity, though it is not easy to define the precise mechanism, which may involve the release of hormones.

New cells are produced in all these regions by a method of cell division called MITOSIS (also referred to as karyokinesis). Reference has already been made to the fact that all the inherited characters are borne on the chromosomes. Since the new individual is originally a single cell produced by the union of the germ-cells, it means that development and expression of the genetic characters (i.e., the whole make-up of the plant) can be attained throughout the individual only if these characters are distributed to all the cells. Thus the nuclear material must be divided equally both qualitatively and quantitatively at each division so that the nuclear constitution is maintained in all the daughter cells. It will be seen that the process of mitosis provides for this.

Cells in the meristematic regions are characterised by the possession of thin walls, dense cytoplasm without vacuoles (though behind the meristematic regions we shall find cells still dividing long after vacuoles have appeared) and large nuclei. Though frequently referred to as resting nuclei when not dividing, a better term is " interphase nucleus ". In this condition the nucleus appears as a fine network of chromatin with denser granules in the meshes and one or more nucleoli enclosed within the nuclear membrane. The clear material within the network is called the nuclear sap.

The whole process of cell division is initiated by the nucleus, and its onset is indicated by the fact that the chromatin network begins to assume a more distinct form, due to the shortening and thickening of the strands. These eventually appear as a single spiral thread which is composed of the chromosomes linked end to end and more or less recognisable. As the spiral shortens and thickens, the individual chromosomes become more distinct. As this happens, the nucleolus gradually disappears, and it is suggested that this body is a reserve of nucleic acid which is taken up by the chromosomes, this being one reason why they stain more obviously during cell division. At the same time the nuclear membrane disappears, and by this time the separate chromosomes can be recognised, each showing a longitudinal split dividing it into two chromatids. During this stage the chromosomes may assume a distinctly beaded appearance, the segments being referred to as chromomeres. When the individual chromosomes are examined it can be seen that each is duplicated and that many of them have a characteristic shape in a particular species. At one point in each chromosome there is a region which is relatively clear and known as the centromere. This also divides when the chromatids are formed. In mitosis there is no special association between the members of a chromosome pair.

The changes so far described constitute **prophase**, and in the later part of this phase alterations have occurred in the cytoplasm. From a point at each end of the cell there appear what seem to be radiating fibrils in the cytoplasm, forming a sort of cone with its base on the mid-plane of the cell. This double-cone structure is called the spindle, its apices the poles and the widest region the equator. The exact nature of the spindle is difficult to determine, but among other explanations it has been suggested that the fibrillar arrangement is due to linear arrangement of the protein molecules producing a state of tension in the cytoplasmic material. Whatever its nature, it is a constant feature of cell division, and there are some differences in spindle organisation between plants and animals.

The chromosomes now come to lie at the equator of the spindle with the centromeres associated with the spindle fibres. This stage is **metaphase**. Actual division now takes place quite suddenly by the separation of the chromatids of each pair. Because the separation always starts at the centromere, the ends of the chromatids remain longest at the equator, so that they appear to be U or J shaped as they move away towards the poles. This is **anaphase**, and ends when the chromatids have bunched together at each pole. Here they undergo a series of changes which are broadly the reverse of those which were seen in prophase and which result in the formation of the daughter nuclei. This is the **telophase** stage, and when it is completed two daughter nuclei have been formed resembling the original nucleus in all respects.

As telophase proceeds the spindle becomes less distinct and gradually disappears, but along the equator small granules of material are being deposited to produce a cell plate. This is the beginning of the new cell wall and gives rise to the middle lamella (which is of a pectin nature), but each new protoplast builds its own wall inside the lamella. It should be emphasised, however, that the wall remains penetrated by fine protoplasmic threads, the plasmodesma, and these maintain continuity between the two protoplasts. As cell after cell is formed by this method of division, it is likely that for some time at any rate many of the protoplasts remain in direct communication. Subsequent differentiation is likely to interrupt this, and it is often suggested that complete rupture of all the protoplasmic connections of a cell is one of the causes of the death of the cytoplasm. The individual nature of the cell wall is shown by the fact that if the middle lamella is dissolved away by chemical action it is possible to isolate individual cells—a very useful method of examining cell structure.

As a result of mitosis, therefore, two cells have been produced and will go on to repeat the process a number of times. The important thing is that each daughter cell has the same chromosome constitution, and therefore the same genetic constitution, as the parent cell and, tracing backwards, as the original fertilised egg-cell. Moreover, this condition will extend to all parts of the adult plant except the germ-cells, so that whenever a new individual is formed by separation of a vegetative structure, the new plant will have *all* the characters of the parent. This is a matter of the utmost importance in horticulture because it means that so long as a particular plant can be propagated vegetatively its special properties can be maintained for generation after generation, and also it is possible to produce a large number of new plants from a single individual and know that they will retain all the features for which it was originally selected. Such a group is called a clone.

Damage to a chromosome during division may prevent development of the cell or result in some peculiarity.

Mitosis is the characteristic division in all the apical meristems, in the cambium and cork cambium and in the embryonic structures. It can be followed very well in root apices, especially those of some Monocotyledons.

In some of the lower plants, and frequently in degenerating tissues in the higher plants, the mitotic mechanism is absent and the nucleus will divide by what is really simple constriction. This of course will often result in the formation of two unequal structures, and is called AMITOSIS.

Meiosis

It will be appreciated from what has been said that the process of mitosis maintains the diploid or $2n$ chromosome number in all the resulting cells. If this process continued unchanged in the development of the mating cells or gametes, the fusion of the latter in fertilisation would lead to the formation of a new individual with $2 \times 2n$ chromosomes, and in turn its gametes would produce offspring with $8n$ chromosomes and so on. This would give rise in a very short time to an impossible situation, although in plants at any rate the lower multiplications are quite frequent and such polyploids are to be found among many of our cultivated plants, where one of the effects easily noticed is that they are much larger than their wild diploid relatives.

However, the normal procedure involves a mechanism which prevents this duplication, and at some stage in the life-history there is a **reduction division** or **meiosis**, as a result of which cells are produced with a **haploid** or *n* complement of chromosomes. The exact stage at which this occurs varies in different groups of plants. In some Algae and Fungi it occurs at the germination of the zygote (the individual produced by fertilisation), so that the individual spends its life in the haploid condition. In the brown seaweed *Fucus* meiosis occurs in the early stages of formation of the germ cells, a condition which is unusual in plants. In the Bryophyta, Pteridophyta and Spermatophyta it precedes the formation of the spore— the asexual reproductive body. The interesting feature of this is that in Bryophyta it means that the familiar Moss or Liverwort plant is haploid, in Pteridophyta the typical Fern plant is diploid but produces a haploid prothallus, whilst in Spermatophyta the pollen grain and broadly speaking the embryo sac are haploid. In all these cases the gametes are finally produced by mitosis from special structures in the haploid gametophyte (a condition very much simplified in the Spermatophyta).

The stages of meiosis or the reduction division can be followed very well in the development of the Fern spore or the Angiosperm pollen-grain.

The spore mother cell is itself produced by a series of mitotic changes from the archesporial tissue in the sporangium or pollen sac. It then undergoes some preliminary stages which are not unlike those of mitosis, but differences soon become apparent and the prophase is a very complex stage.

The chromatin material becomes aggregated into an elongated beaded thread, which at a very early stage separates into the diploid number of chromosomes. This is the **leptotene** stage. The similar or **homologous** chromosomes then associate in their pairs and become shorter and thicker as the nucleolus disappears. The centromeres become apparent during the **pachytene** stage, and this is followed by the **diplotene**, which each chromosome divides longitudinally into two chromatids. At this stage a complex situation develops, in that the chromatid pairs become twisted round one another

in varying degrees, forming what are known as **chiasmata.**
Because the chromosome material is not rigid, but of a sticky
nature, it frequently happens that exchange of chromatid
sections may occur between chiasmata or from a chiasma to the
free end of a chromatid. Such a condition is known as
crossing over, and it means that when the chromosomes (or it
may be more illuminating to say chromatid pairs) eventually
separate, their structure differs from the original condition,
and this has a very important bearing on genetics. It seems
to be the case that the exchange is always between chromatids,
and not between whole chromosomes. The chromatids which
separate are superficially like the original ones, and there is a
quantitative balance of material in the normal way. (Anomalous changes may give freak results but it has been suggested
recently that the significance of the chiasmata may not be
fully known.)

Further shortening and thickening of the chromosomes now
occur during **diakinesis,** which is the final stage of prophase.

As in mitosis, a spindle has formed and the chromosome
pairs go into metaphase on the equator.

When separation occurs at anaphase, members of each
chromosome pair pass to opposite poles, so that a single set of
chromosomes gathers at each pole, the chromatids still remaining in association with one another.

There is now a brief interval, after which—in plants, at any
rate—both daughter nuclei undergo a second division, which
superficially resembles mitosis, in that the chromatids now
separate along a spindle at right angles to the original one, so
that four cells are finally produced. Each of these possesses a
haploid nucleus, and though from the nature of the first meiotic
division we should expect two of the nuclei to differ from the
other two, the crossing over between chromatids means, in fact,
that at the second-stage division the nuclei which are formed
are not quite identical, so that this second-stage division is not
a true mitotic one. So the gametes which will eventually be
produced will have different chromosome constitution.

It may be remarked that in the formation of animal eggs
only one cell of the four is ever properly formed, whilst in the
embryo-sac formation in Spermatophytes three of the cells
produced usually degenerate.

The results of this type of division are rather complex.
Evidently in groups such as the Bryophyta the haploid nucleus
possesses all the genetic factors required to produce the plant,
because it has already been pointed out that the ordinary Moss
plant is haploid throughout. What the evolutionary effects of

diploidy have been is not easy to see, but it seems that it is a more advanced condition, since it is the rule in all the higher organisms.

It is probably sufficient to say here the meiosis ensures the maintenance of the diploid condition at fertilisation and also provides for the recombination of genetic factors during that process—an event which is the starting point of the mechanism of evolutionary selection.

(ii) Tissues

Although at the apical meristems the cells are all very much alike, it is not long before there is an indication of the future distribution of the tissues. Thus the surface layer is well defined, and divides only by walls at right angles to the surface. This layer is called the dermatogen and gives rise to the whole covering layer of the plant—the epidermis. The inner layers of cells become recognisable as a central core called the plerome and a surrounding region (up to the dermatogen) which is the periblem. These layers are not always well marked and quite often they are distinguished by their products rather than by their appearance. This organisation is essentially the same for stem and root (the leaf being merely a lateral fold from the stem apex), but the latter has an extra layer, called the calyptrogen, which gives rise to the root-cap. In general the periblem produces the adult region known as the cortex, whilst the plerome gives rise to the stele, which embodies in particular the conducting or vascular system of the plant.

In order to study the structure and distribution of the tissues in plant organs several methods are used. One of the best known is to cut thin sections of the organ concerned in various planes; transverse sections at right angles to the long axis and longitudinal sections parallel to the long axis. Harder tissues may be cut with an ordinary hand razor or, if very woody, by a heavy razor on an instrument called a microtome. Small organs and very soft delicate tissues are embedded in wax or some other material to hold them firm and cut in serial sections on another type of microtome. The tissues are then stained and mounted on glass slides for microscopic examination. Such a procedure gives us a very fair idea of the mechanical arrangement and general distribution of the tissues, but it must be borne in mind that the tissues are now dead and we can learn little of the relationships of the parts and their functional behaviour. In many cases we are really examining a system of empty cell walls—a kind of skeletal arrangement. Sometimes, however, by careful methods of fixation it is possible to retain the cytoplasm and get some idea of its structure.

But the method of sectioning does not always demonstrate satisfactorily the structure of the individual cell, and this is particularly the case in longitudinal sections. It is not easy to cut a section only one cell thick, and even then much may be missed.

So for this purpose a very useful method is to break down the middle lamella between individual cells so that the latter fall away from one another and can be examined separately. This process is called maceration, and is very useful for observing the actual size, etc., of individual cells. Obviously it is of no use for studying tissue arrangement.

When the plant organs are studied by the various methods described, it is found that although the distribution of the tissues is different in the various organs, the individual elements are very similar for any one tissue, a statement which to a considerable extent holds good beyond the Flowering Plant.

It is proposed to study representative tissues by examining the structure of an ordinary herbaceous stem as seen in transverse section and by maceration.

(iii) The Internal Structure of the Stem

If a longitudinal section of the tip of a stem is examined it is possible to follow the gradual differentiation of the tissues from the meristem into the adult region.

In most Flowering Plants the organisation is fairly constant and shows a gradual sequence of embryonic leaves arising as small outgrowths from the surface layer near the apex in a sequence which has been discussed earlier. In the stem itself, and later in the leaf-base, it can be seen that there is a gradual change in the shape and distribution of the cells. Behind the more or less hexagonal (in section) cells of the apical meristem their successors become elongated, though still continuing to divide. The surface cells constituting the dermatogen divide only by walls at right angles to the surface, and, as they grow rapidly, folds are formed which are the potential leaves or leaf primordia.

The inner cells elongate and enlarge rapidly, presumably due to the intake of water, so that cytoplasm is not formed quickly enough to fill the cells, and thus vacuoles appear. Gradually these cells lose their meristematic properties and become adult. Meanwhile the wall becomes thicker due to deposition by the cytoplasm of a carbohydrate called cellulose, which is the basis of all plant-wall structure.

However, not all the cells differentiate in this way, and in that group of plants known as Dicotyledons a system of cells roughly forming a cylinder and continuous with the apical meristem remain meristematic but become elongated. These cells are called procambial cells, and at first they divide in two longitudinal planes (i.e. parallel to the long axis of the stem). The products of these divisions differentiate to produce the first conducting or vascular tissues, and these are formed in a manner which will be described later. The procambium is the outermost region of a primary tissue mass called the plerome which fills the central core of the stem. The region between the plerome and the dermatogen is the periblem. In Monocotyledons the formation and differentiation of the procambial strands are rather different.

In the young stem the groups of procambium differentiate into systems called vascular bundles, which in some cases are separate, but in many others remain continuous, so that a ring of vascular tissue is developed. The formation of separate vascular bundles is quite frequent, though by no means invariable, in herbaceous types, but in woody plants there is almost always a cylinder of vascular tissue from the earliest growth of the bud. It is important to draw attention to this point because in the Sunflower and one or two other forms generally chosen to demonstrate plant anatomy the method by which a complete vascular cylinder is developed in the older plant is rather different from the course of events in the woody plant.

The general condition derived by the differentiation of the procambium is often called the primary condition, but in Dicotyledons it is often difficult to define, as will be explained shortly.

Fig. 16 shows a transverse section through the stem of the Sunflower (*Helianthus annuus*). Although this plant is often used as a type, it must be realised that no one plant can serve as an example to be closely followed by all other species, but the main features are general, and the detailed distribution does not affect the understanding of the general principles of stem anatomy. The same argument applies to the other regions of the plant.

The principal tissues will now be discussed in more detail.

The Epidermis

If one of the transverse sections is examined in detail it will be found that the outer layer consists of cells fitting closely together and more or less rectangular in outline. This layer

is the epidermis, and is usually covered by a non-cellular deposit
of fatty material which is known as cuticle, and which varies
a great deal in thickness. The epidermal cells have, at least
when young, a cellulose wall and are not separated by intercellular spaces. This is an important protection against loss
of water by evaporation. However, it is found that among the

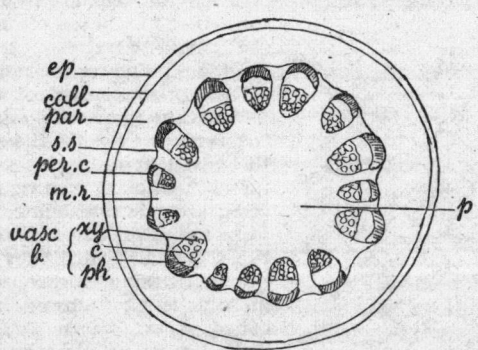

FIG. 16.—Transverse Section of Stem of Sunflower. Diagrammatic.

ep. epidermis, *coll.* collenchyma, *m.r.* medullary ray, *par.* parenchyma, *per. c.* pericyclic cap, *s.s.* starch sheath, *vasc. b.* vascular
bundle, *xy.* xylem, *c.* cambium, *ph.* phloem.

epidermal cells in the young plant there are special openings
called stomata, which permit of gas exchange. The operation
and structure of these will be discussed when the leaf is described. The epidermis may be smooth or may develop various
kinds of epidermal hairs, whilst in more woody stems prickles
are quite common (e.g. Blackberry and Rose).

As the plant grows older the epidermis changes, and in particular the cell wall becomes harder, due to further deposition
and to changes in the material of the wall. Frequently the cells
die as new tissues develop below.

The Cortex (Fig. 17).

The layer of cells immediately within the epidermis is the
exodermis, and it may have special characters or may resemble
adjoining layers. It is the outermost layer of the region called
the cortex, and in the Sunflower there is a zone two or three
cells deep below the epidermis which consists of cells called
collenchyma. These cells are roughly cylindrical, with flat

end walls and large vacuoles. Their peculiarity is that they have extra deposits of cellulose on certain parts of the wall—those adjacent to intercellular spaces. This extra thickening gives added strength with flexibility to the structure in which such cells occur (it is common in leaves). Collenchyma may appear as a complete band of cells or in blocks, the latter being frequent in angled stems. Collenchyma cells are living, and form part of what is often called the ground tissue.

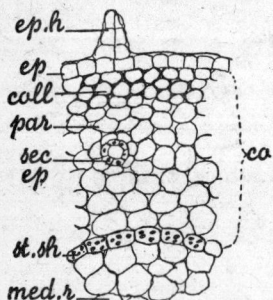

FIG. 17.—Transverse Section of Stem of Sunflower. Detail of Cortex.

ep. epidermis, *ep.h.* epidermal hair, *med.r.* medullary ray, *coll.* collenchyma, *par.* parenchyma, *sec.ep.* secretory epithelium, *st.sh.* starch sheath, *co.* cortex.

In the Sunflower the remainder of the cortex consists of what is perhaps the most widely distributed type of tissue in the plant—the **parenchyma.** This tissue is found in all plant organs, and the cells are always living. Their form varies greatly, according to the region in which they occur, but in the stem cortex they are more or less cylindrical, like the collenchyma. When young the walls are of cellulose and often quite thin. Large vacuoles are present, and not infrequently the outer parenchyma cells possess chloroplasts. This is generally the case in young herbaceous plants. The innermost layer of the cortex is not broken by intercellular spaces and the cells frequently possess starch grains even when other cortical cells are without. For this reason it is referred to as the starch sheath, although technically it is the **endodermis.** In different organs it shows a considerable range of modification, and is particularly important in the root.

Many plants have another tissue in the cortex but this is not the case with Sunflower. This is called sclerenchyma, and consists of dead cells which in many cases have been subjected to pressure so that they are elongated and pointed. The walls become woody and the cells are called fibres. They act as strengthening tissue.

When the various cortical cells are examined by maceration, the walls are frequently found to have thin places, and in many cases these correspond on adjacent cells. These thin places are called pits, and very probably represent regions in which cytoplasmic connections extend between the two cells, so that the successive deposits of wall material laid down by the cytoplasm were interrupted. It must be realised that all additional material deposited on the cell wall is produced by the cytoplasm and is therefore on the inside of the original wall.

It is not easy to follow or to understand the precise changes and developments in wall structure. The original wall is called the primary wall, but it is almost impossible to say that at such and such a point it becomes secondary. Moreover, there is sometimes a tendency to misuse the term thickening—a wall frequently changes in chemical nature without further deposition of material. Thus the sclerenchyma fibres of which we have spoken become woody due to the formation of lignin, which is not an actual deposit, but is formed by changes which gradually take place in material already present in the wall, involving in particular the loss of water. Lignification is often not completed until long after the death of the cell.

Various other structures may occur in the cortex of individual species. Thus in the Sunflower resin ducts are present. These are tubes enclosed by a layer of secretory cells the products of which pass along the tubes, though for what function is not always clear.

In other plants, like Dandelion, Sow-thistle and the Spurges, there are long tubes called laticiferous vessels which produce a milky fluid called latex. A similar fluid is collected to produce rubber from trees of the genus *Hevea*.

It will be recalled that the innermost layer of the cortex was the starch sheath, and within this lies a region called the **stele**. In most of the examples encountered the stele shows an outer zone of vascular tissue (which, as already explained, may or may not be in separate vascular bundles) and a central mass of parenchyma which constitutes the pith. In a number of cases the rapid expansion of the stem may rupture the pith, and the middle of the stem is then hollow—a condition frequently found in herbaceous stems.

The Stele

Immediately within the starch sheath is a layer of cells called the pericycle and of varying depth. In many cases they are parenchymatous, and are not then particularly noticeable. But in some cases, including the Sunflower, the pericyclic cells immediately outside each vascular bundle become sclerenchymatous and form a protective cap. In the case of Vegetable Marrow there is a complete ring of pericyclic sclerenchyma, and it is not associated with the vascular bundles.

It will be found that in the root the pericycle plays a much more varied role in activities of the organ.

The Vascular Tissues

The conducting elements of the plant are grouped into two kinds of tissue, which are called phloem and xylem. It may be said straight away that it is generally considered that the phloem is responsible for the transport of organic substances synthesised in the plant, whilst the xylem provides the channels along which watery solutions of mineral salts flow from the regions of absorption in the root to all parts of the plant.

These tissues are associated in different ways, but in general they are found in the stem in bundles (Fig. 18), in which the phloem is directed towards the outside (though in some cases—e.g. Vegetable Marrow—there is a second zone of phloem on the inner surface of the xylem). Even where the original meristematic tissue produced a complete cylinder of differentiation the vascular tissue is more developed in some places than others, and these regions correspond to the separate bundles of the Sunflower and similar types.

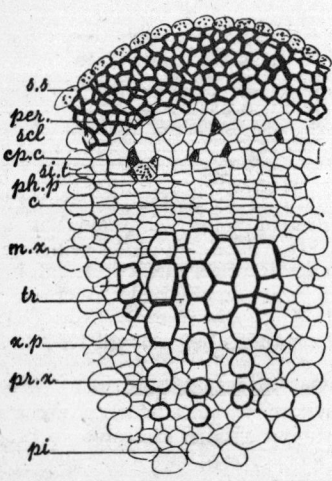

FIG. 18.—Detail of Vascular Bundle of Sunflower Stem.

c. cambium, *cp.c.* companion cell, *m.x.* metaxylem vessel, *per.scl.* pericyclic sclerenchyma, *ph.p.* phloem parenchyma, *pr.x.* protoxylem vessel, *pi.* pith, *s.s.* starch sheath, *si.t.* sieve-tube, *tr.* tracheid, *x.p.* xylem parenchyma.

The majority of these bundles are continuous from stem to leaf, but in the stem they tend to join, so that there is a network from which the leaf-bundles or leaf-traces depart at intervals. It must be understood that there is a very close association between the development of the vascular system of the leaf and that of the stem, though much of the vascular tissue in the leaf does not pass through the leaf-base into the stem.

In most herbaceous Dicotyledons the vascular tissue is present in a single ring or as a single ring of bundles, but sometimes a second ring is present, whilst in some aquatic stems at least the vascular tissue may be reduced to a single central strand.

Development

The vascular tissues develop from the meristematic layer known as the procambium or from its derivative layer, the cambium. Tissues developed in the first way are called primary, whilst those originating from the cambium are secondary. There is some confusion about these terms, but the above derivation has been recognised. Since the vascular elements provide a system of tubular structures, it follows that their development must be in a linear, longitudinal series, though junctions may occur. The exact procedure varies with different types of element, but one feature which becomes clear as we go along is that whilst the primary tissues show no particular radial orientation, the tissues produced from the cambium do appear in radial series when viewed in transverse section. This is due to the fact that in the cambium most of the divisions take place in the tangential plane, so that new cells appear to the inside or outside of the actual cambial ring. Later differentiation may distort this appearance, but it remains very obvious in the wood.

When the cambium develops early most of the vascular tissues show this radial appearance, whilst in Monocotyledons and some herbaceous aquatics where cambium does not appear the tissues show no radial disposition at all. On the other hand, in woody stems this arrangement is very marked.

When the bundles are separate the cells lying between them constitute the medullary rays, and even where there is a cylinder of tissue these rays appear and are maintained in the woody condition.

The Xylem

The xylem is that part of the vascular tissues which forms the wood, and it possesses certain characteristic conducting elements known as vessels and tracheids. In addition, fibres are frequently present together with parenchyma, but these elements do not differ materially from their condition in other parts of the plant.

(a) Vessels. These are peculiar to Flowering Plants and represent a series of cells from which the end walls have disappeared. The manner in which this has occurred is not too clear, but it is usually believed to be due to pressure within the element in its young state, especially if this wall is stretched by rapid lateral expansion of the element. At first the cells which later form the vessel are thin-walled and living and the cytoplasm forms a thin lining within the wall. Even when the end walls disappear the cytoplasm is not immediately lost, and during this time it deposits further material on the wall. This forms the secondary wall. The way in which this material is deposited is affected by the activities of the stem. If the latter is still elongating, the developing vessel is stretched by the pull of the surrounding tissues and the new wall material is not deposited evenly, but in annular bands or, more commonly, in a spiral band. As growth slows down the spiral becomes closer, and in the non-elongating parts of the stem the new material is deposited much more evenly, but is absent from the regions which will become the pits.

New elements are constantly being differentiated behind the apex in continuity with the older ones and in association with the new leaves. Usually the first xylem elements to be distinguished are those on the inner surface of the leaf-trace. The elements produced during elongation are usually called **protoxylem**, and these show annular and spiral thickening of the walls. Those which have developed after elongation has ceased form the **metaxylem**, and these are pitted. It must be emphasised that in Dicotyledons, at least among the Flowering Plants, the new elements formed behind the growing apex are in continuity with others produced by the cambium, so that even in large trees connection is maintained from primary root up to developing leaf through the new xylem elements.

The xylem vessels are not all equal in the degree of lateral expansion, so that we may get very wide ones and very narrow ones, a point which will be further discussed later. Moreover, the expansion of the wider elements may result in the crushing of surrounding cells, leading to their elongation and early death. It is in this way that many of the xylem fibres are produced.

(b) **Tracheids.** Not all the differentiating xylem elements undergo loss of the end walls. In a number of cases the element undergoes changes similar to those found in a vessel, but the end wall is not broken down, and in such cases the end wall is often found to be oblique. These elements show the same type of secondary thickening as the vessels, so that there are annular, spiral and pitted tracheids. Their distribution is uneven and they are more frequent in some species than in others. The first-formed xylem elements are often tracheids, whilst the vascular terminations in leaves consist of one or more tracheids.

In the Ferns and practically all Gymnosperms there are no vessels, and tracheids then form the conducting tissues of the xylem.

As the elements differentiate and become older their walls become woody due to the formation of lignin, which is a complex material allied to the sugars and exceedingly durable. By the time this has developed the elements are dead, but they may continue to conduct for many years until blocked by deposits of tannin and other materials and by the protrusion of parenchyma cells through the pits, an occurrence which produces a number of blister-like structures called tyloses. Fig. 19 shows some individual vessel segments and tracheids.

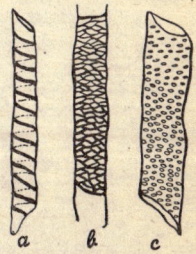

FIG. 19.—Types of Vessel Segment.

a. spiral, *b.* reticulate, *c.* pitted.

Besides the actual conducting elements, fibres are present, and these have already been mentioned. In addition, parenchyma is rarely absent from the xylem, and though it becomes lignified it remains alive for a long time and may be responsible for the movement of soluble materials laterally through the xylem.

Due to the woodiness of the constituent tissues, the xylem does not become crushed, and so forms a large part of the older stem. This becomes even more obvious in the woody plant, where it forms a very large proportion of the whole organism.

The Phloem

The other main conducting tissue is the phloem, and here it will be found that the functional life of the principal elements is much shorter than in the case of the xylem. This may be due to the fact that they do not undergo lignification, and are thus subject to crushing.

In most Flowering Plants the phloem is in the same bundle as the xylem and outside it. Such an arrangement is described as conjoint and collateral. Where a second group of phloem elements appears on the inner face of the xylem the bundle becomes bicollateral, and such a condition is most easily seen in some of the climbing plants, such as Cucumber, White Bryony, etc.

The Sieve Tube

The actual conducting element of the phloem is the sieve tube, and the general appearance of some of these elements is shown in Fig. 20. As with xylem vessels, the sieve tubes develop in longitudinal series, but there is never a complete breakdown of the end wall. What does happen is that the end wall becomes perforated (perhaps along the track of the cytoplasmic connections) so that there is free connection from segment to segment. This perforated area is called the sieve plate, and though normally there is only one sieve plate, some plants with rather sloping end walls to the sieve-tube segments

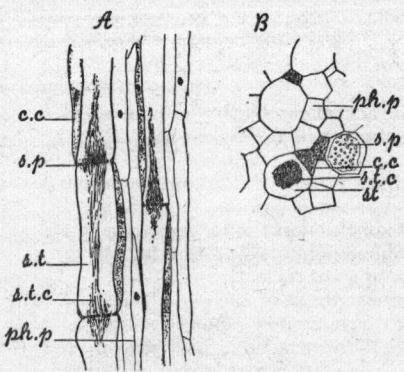

FIG. 20.—Sieve-tubes.

A. Longitudinal view. B. Transverse view.

c.c. companion cell, *ph.p,* phloem parenchyma, *s.p.* sieve-plate, *s.t.* sieve-tube, *s.t.c.* sieve-tube contents.

have two or more areas forming a compound sieve plate. What seems to be less easily explained is the presence in quite a number of plants of small sieve areas on the lateral walls (usually the radial ones), and indeed in the Ferns and Gymnosperms this is universal condition.

The earliest sieve tubes are again developed from the procambium, whilst later elements are differentiated from the cambium. Unlike the condition in the xylem, there is little structural difference between primary and secondary phloem or between protophloem and metaphloem, since we do not get the wall-changes which produced this distinction in the xylem. As the sieve tube differentiates it expands laterally in varying degrees, but in general they do not reach the width attained by xylem vessels. During the differentiation the cytoplasm becomes vacuolated and restricted to a lining round the wall, but the sieve-tube segment remains alive for a long time, and indeed it is quite probable that the element retains its conducting function only whilst the cytoplasm is present. Ultimately the element dies, and usually by this time the sieve plate has become blocked by material that is presumably deposited from the substances constantly passing through the tube. Even in a herbaceous stem the earliest-formed tubes may show signs of crushing, due to the expansion of later elements. Further reference will be made to the condition in woody stems.

Companion Cells

One of the most remarkable features of the sieve tube in the Flowering Plant is the fact that it never differentiates alone, but is accompanied by a cell formed as a result of a longitudinal division in the radial plane soon after the sieve-tube initial has itself been formed. The two cells so produced do not expand equally, and it is the wider one which becomes the sieve tube. The other is called the **companion cell,** and remains narrow with rather dense cytoplasmic contents. Very often, however, it undergoes a series of transverse divisions, so that a single sieve-tube segment may be accompanied by a row of companion cells. The actual function of the companion cell is not clear and it undergoes little specialisation, though it has been claimed that the companion cell may vacuolate. The companion cell does not appear to have any special storage functions, and it must be said that most of the possible roles attributed to it are speculative. Companion cells are absent from the other groups of vascular plants.

The phloem may contain other tissues which are non-conducting. Parenchyma is of widespread occurrence, and in woody stems patches of sclerenchyma fibres are often to be found.

It may be pointed out that air-spaces are rarely found between the elements in the vascular bundles, but there may be some development of such spaces in the parenchyma of the woody vascular cylinder, especially adjacent to the medullary rays.

Pith

In most herbaceous dicotyledons the remainder of the stele is occupied by parenchyma, the cells of which are often very large. This central zone is called the pith, but in a number of cases, possibly because of the tension due to rapid growth, the pith becomes ruptured and a pith cavity forms. Where the pith-cells persist they frequently become lignified.

Between the bundles parenchyma extends from the pith to the cortex, and these strips of tissue are called medullary rays. They do not seem to have any important function in the herbaceous stem, but more will be heard of them in the woody structure.

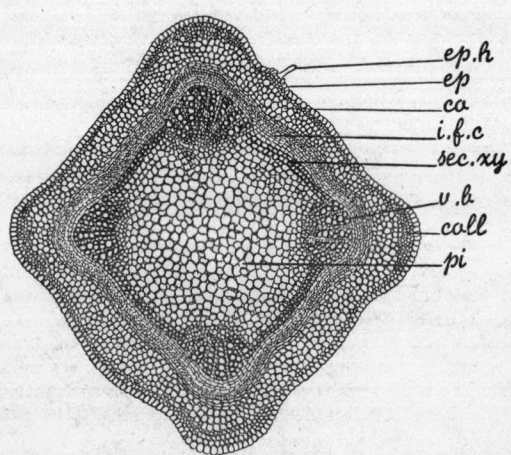

FIG. 21.—Transverse Section of Stem of Deadnettle.

co. cortex, *ep*. epidermis, *ep.h.* epidermal hair, *i.f.c.* interfascicular cambium, *pi*. pith, *sec.xy.* secondary xylem, *v.b.* vascular bundle, *coll*. collenchyma.

In this brief review of the structural features of an ordinary herbaceous stem it has been stressed that one cannot very well use the term " typical stem ", because differences will be observed from species to species. On the other hand, major differences in organisation usually occur only when there is some ecological influence. Thus in aquatic stems the amount of vascular tissue, particularly xylem, is greatly reduced, though it cannot safely be inferred that this is directly associated with the decreased importance of water conduction. In such stems it is frequently the case that there is a simple ring of xylem vessels in the middle of the stem surrounded by the phloem,

which is in turn enclosed by the pericycle and a well-developed
endodermis.   Thus there is no separation of the vascular

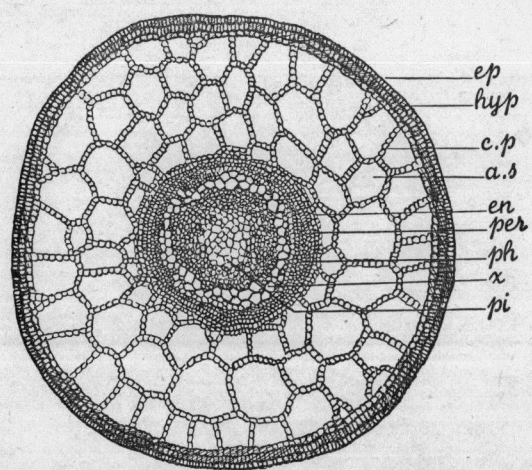

FIG. 22.—Transverse Section of Hippuris Stem, an Aquatic-Type.

a.s. air space, *c.p.* cortical parenchyma, *en.* endodermis, *ep.*
epidermis, *hyp.* hypodermis, *per.* pericycle, *pi.* pith, *ph.* phloem, *x.*
xylem.

tissue into bundles and often there is no secondary tissue.
Some examples of Dictoyledon stem structure are shown in
Figs. 21 and 22.

Monocotyledons

When the stem structure of the Monocotyledons is considered, some well-marked differences are to be found.   Fig.
23A shows the general distribution of the tissues in the stem
of Maize, and this can be regarded as representative of many
herbaceous Monocotyledons.   Fig. 23B illustrates another
common form as seen in the flowering-stalk of Wheat.

In these stems the epidermis is well defined and in the older
stems frequently become heavily lignified.   Within this tissues
are not differentiated into cortex and pith, though there is
often a very definite hypodermis which also becomes lignified,
at least in many of the Grasses.   As will be seen in the diagram
of the Wheat stem, it is quite usual to find the stems with a

large central cavity, though this is interrupted at the nodes, which are solid.

Probably the most familiar feature of the Monocotyledon

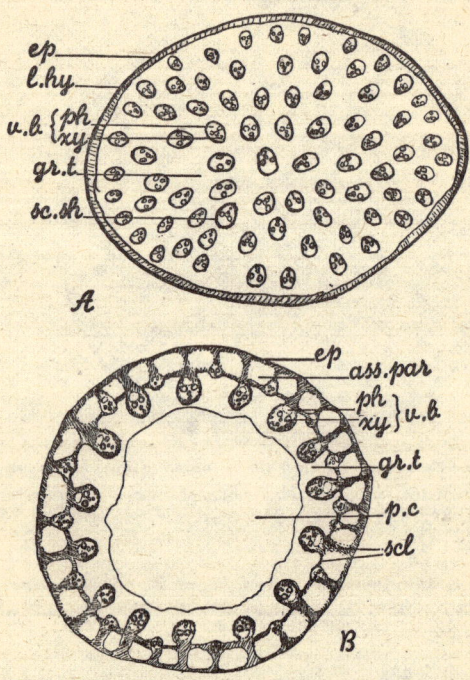

Fig. 23.—Monocotyledon Stems.

A. Maize.   B. Wheat.

ass. par. assimilating parenchyma, *ep.* epidermis, *gr.t.* ground tissue, *l.hy.* lignified hypodermis, *p.c.* pith cavity, *sc.sh.* sclerenchyma sheath, *scl.* sclerenchyma, *v.b.* vascular bundle, *ph.* phloem, *xy.* xylem.

stem is the scattered arrangement of the vascular bundles, though this is more restricted in the hollow stems. The arrangement is not really haphazard, but is due to the large number of leaf-traces entering at each node. These bundles follow a complex course and bend steeply into the stem when they first enter, though later there is a good deal of fusion or anastomosis.

Although the bundles contain both phloem and xylem as before, there is never any cambium, so that all the tissues are primary. Such bundles are said to be closed. Phloem parenchyma is often very restricted, and the tissue between phoem and xylem is called conjunctive parenchyma. Very

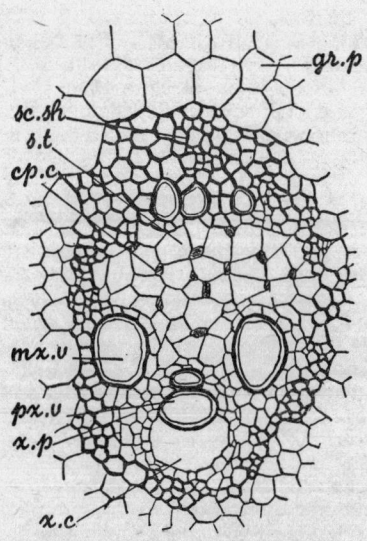

FIG. 24.—Detail of the Vascular Bundle of Maize.

cp.c. companion cell, *gr.p.* ground tissue parenchyma, *mx.v.* metaxylem vessel, *px.v.* protoxylem vessel, *sc.sh.* sclerenchyma sheath, *s.t.* sieve-tube, *x.p.* xylem parenchyma, *x.c.* xylem cavity.

often the bundles are enclosed by a sclerenchyma sheath. Such a bundle is shown in Fig. 24. Because of the absence of cambium, normal secondary development does not take place, so that there is no later cylinder of vascular tissue. On the other hand, many Monocotyledon stems, and particularly the Grasses, undergo a great deal of lignification of the ground tissues, so that a hard but rather brittle organ is formed. This is easily seen if the old stems of grasses and the ripe stems of Wheat, Barley, etc., are examined. Some of the fibrous strands are very hard and can easily cut one's finger. An extreme example of this development is seen in the Bamboos.

The individual tissues are similar in structure to those already described, and no further detail need be given.

One or two interesting similarities with certain Dicotyledon stems may be mentioned. The climbing plant Black Bryony has a vascular system which in many details closely resembles that of Vegetable Marrow and White Bryony, though, unlike these plants, there is no cambium. This is an interesting example where similarity of habit or functional development (often accompanied by environmental influence) has produced a structural relationship more close than that which exists between species which have a classificatory relationship but different environmental conditions. One or two other examples will be quoted as they occur.

(iv) The Further Development of the Stem

So far discussion has been limited to the herbaceous condition of the stem, and it has been pointed out that it is not easy to mark the boundary between primary and secondary conditions. Even in the herbaceous stem the tangential divisions of the cambium add new elements to both xylem and phloem, with the result that the girth of the stem gradually increases. Usually the production of new xylem proceeds faster than that of phloem, but in both cases the differentiation of the tissues lags behind formation, so that in rapidly expanding stems the zone of immature tissue is often extensive.

Generally this secondary thickening, as it is called, is ended by the death of the herbaceous plant, but even so an old Wallflower or Sunflower stem will show a complete cylinder of xylem of considerable thickness, though the positions of the original bundles are still recognisable. The radial arrangement of the xylem elements is very obvious, and the files of vessels, etc., are interrupted at intervals by new rows of parenchyma which are the secondary medullary rays. The phloem is also increased but this usually results in the crushing of the earlier phloem tissues. A general feature in these old herbaceous stems is the increasing lignification of the parenchyma with advancing age. This may be associated with the greater availability of synthesised materials and the loss of water during the summer.

But the real development of woody tissue is seen in trees and shrubs. Fig. 25 is a diagram of a transverse section of a three-year-old Apple twig, and it is quite obvious that there is a considerable proportion of xylem in this stem. Most woody stems show a similar distribution of tissues, though there are detailed differences.

In the case of a herbaceous plant with separate bundles the development of a vascular cylinder is achieved by the return to

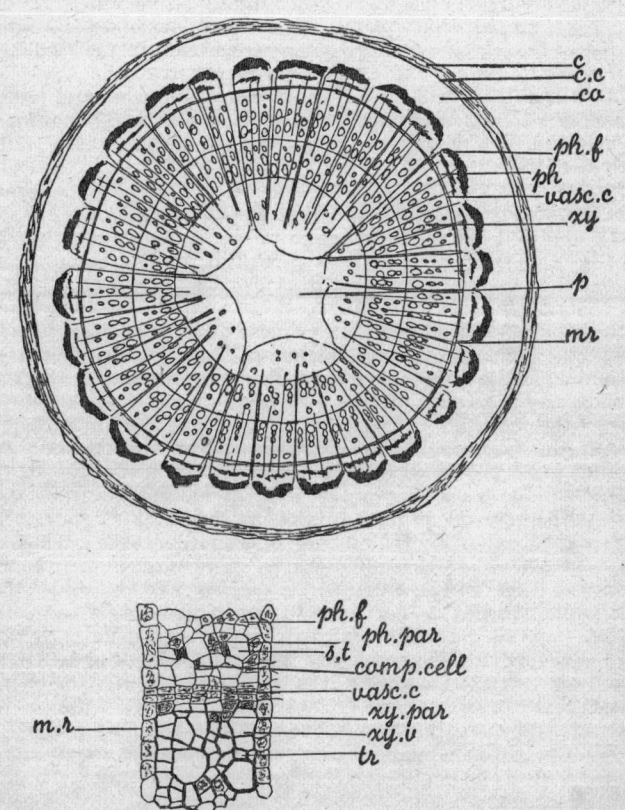

FIG. 25.—A. Transverse Section of Three year old Twig of Apple. Diagrammatic.   B. Detail of Vascular Tissues.

c. cork, *c.c.* cork cambium, *co.* cortex, *comp. cell.* companion cell, *m.r.* medullary ray, *p.* pith, *ph.* phloem, *ph.f.* phloem fibres, *ph.par.* phloem parenchyma, *s.t.* sieve-tube, *tr.* tracheid, *xy.* xylem, *xy.par.* xylem parenchyma, *xy.v.* xylem vessel, *vasc.c.* vascular cambium.

meristematic activity of a zone of tissue linking the cambium of one bundle with that of the next—the so-called interfascicular cambium. In other cases the interfascicular meristematic zone is always present, but only in the older stem does it begin to produce cells which differentiate into vascular tissues.

However, in woody plants this gradual development from a ring of bundles is probably never seen except in the seedling. If a section is cut through the opening bud of the Apple or other tree there will already be a ring of meristematic tissue and very soon there will be a ring of wood. It is true that for a short time the tissue is sufficiently undifferentiated to be soft, so that the young twig is still " herbaceous ". It is obvious, however, that this green, soft stage is very rapidly changed into a woody condition, and that the elongation of the stem is completed before it becomes hard and woody. The length of the elongating zone varies a great deal, and is usually much greater in young plants than in older ones.

As the new cells produced by the cambium differentiate, the rapid formation of xylem pushes out the encircling tissues and the stem increases in thickness. The onset of this new cambial activity in the woody plant is associated with the buds, and it passes down as a wave of growth over the whole tree, the meristem of the new bud being in direct continuity with the hitherto resting cambium of the twig below. In order to keep pace with the increase in thickness of the stem, the cambium cells must occasionally divide radially.

As the new xylem differentiates, its character is determined to a great extent by the amount of water available. Thus in spring, when the tree is full of water, many of the new elements expand vigorously, giving wide vessels and crushing the adjoining tissues so that tracheids and fibres are formed. In trees like Oak this results in the formation of a definite ring of wide elements which constitute the spring wood, but in Apple and others the wide vessels, though decreasing in diameter as the season progresses, are fairly scattered. In Oak the condition is known as ring porous, and in Apple as diffuse porous.

In the earlier stages of wood development the leaves are very young and small, and little water is being lost by transpiration. As the leaves open, this loss of water increases very rapidly, and less is available to expand the differentiating vessels, etc., and so the elements produced in the summer are narrower. Some of the cambial derivatives differentiate into new medullary-ray parenchyma and these rays are common to xylem and phloem. They appear to be channels for aeration.

Phloem is being formed as well as xylem, but here there may

be a difference. Though much of the tissue is undoubtedly produced by the new cambial activity in the spring, there is evidence that some undifferentiated tissue is present throughout the winter, and this differentiates into new sieve tubes and companion cells when growth recommences. In many cases the earlier sieve tubes are wide with large sieve plates, but the later ones are little wider than the cambium elements from which they are derived, and many of these later-formed sieve tubes have two companion cells. Patches of fibre are very frequent in the phloem.

As the leaves become fully expanded the cambial activity slows down, and by July it has usually stopped, though differentiation of the new tissues may go on for some time. The formation of further tissue is not usual, and so by the time that all the wood is fully developed and lignified there is a well-defined zone of wood representing the season's growth. This is called an **annual ring** and, as we have seen, is usually characterised by the formation of wide vessels in spring with narrower ones in the summer. There is *no* autumn wood. Practically all the wood is metaxylem, and is therefore pitted in various degrees. Because the phloem is more easily crushed, the annual production is not so well defined, but in some cases it can be traced.

This seasonal production of vascular tissue follows a general pattern in woody plants, and each year the girth of the stem is increased by an amount of wood of which the extent and even the composition will be influenced by growing conditions, so that besides being able to tell the age of a tree by counting the rings of wood, experts can frequently form some idea of growing conditions and weather cycles at the time of formation of the rings. Thus if there is plenty of water available when the spring elements are expanding one tends to get well-marked zones of wide vessels, whilst a fine summer tends to favour photosynthesis, and a considerable zone of summer wood with narrower and thicker-walled elements appears. On the other hand, a poor growing season will result in a narrower ring altogether.

As the actual production of new cells ceases, differentiation continues, and usually by early autumn all the new xylem and much of the phloem are mature, whilst the cambium appears as a single ring of narrow cells between the xylem and phloem. It remains in this condition until the following spring. Reference has already been made to the fact that some of the phloem remains undifferentiated through the winter.

This pattern of activity is followed in all trees (except

Monocotyledons like the Palms) and bushes, and gradually a dense cylinder of wood is formed which supports the whole plant. It must be emphasised again that the new vascular tissues produced in the first-year twig are continuous over the whole surface of the cambium from the apex of the young shoot to the furthest vascular region of the root. In this way there is a complete conducting system. As far as the phloem is concerned, the current season's tissue is probably the only phloem which is functional, but in the wood the vascular elements continue to conduct for a number of years. Gradually they seem to become blocked by resins, tannins, etc., and by the expansion of xylem parenchyma to form tyloses in the cavity of the vessel. This blocked wood is the **heart wood**, whilst the conducting tissue is the sap wood. The difference between the two regions is very well illustrated by a transverse cut across an older branch or trunk of the Laburnum because in this species the heart wood is very dark. When trees are used for timber the sap wood is less valuable, because if it is used too soon it will contract as the water dries out, and this will make it crack. It improves if allowed to dry out slowly, but of course the heart wood is the more satisfactory.

Cork

So far we have discussed the development of the vascular tissues in perennial woody plants, but there are changes which occur outside this region.

As the stem expands the outer layer is subject to great stress and the cortical cells tend to be crushed by radial expansion and to be distorted by lateral pulling. Whether this is the cause or not, a layer of cells, usually in the outer region of the cortex and occasionally the epidermis itself, becomes meristematic again and divides to form new cells which are mainly towards the outside of the stem. These cells, like other cambial derivatives, appear in radial files, but they do not live very long and give rise to an outer protective layer which is known as cork. This cork or periderm becomes impregnated with various organic materials, particularly a fatty substance called **suberin**, whilst other substances, probably waste products, fill the lumen of the cells. The cork is thus to a large extent waterproof. Cork is a seasonal production, and the manner of its formation determines the appearance of the outer covering of the older parts of the tree.

Most young twigs have a smooth layer of cork, but it remains smooth only if the cork cambium or **phellogen** is present as a continuous sheet just below the surface. This is the condition found in Birch and Cherry, and in these trees large flat strips of cork can be peeled off. Beech and Sycamore also possess fairly smooth barks because the cambium develops as an even layer, but in trees like Oak, Ash, Elm and Pine the cork cambium, after being active for some time, ceases its activity and a new layer develops at a deeper level. Not only that, but these new layers may arise in separate patches, with the result that the new cork is produced unevenly and tensions are set up which give the grooved bark of these trees.

In some cases the cork cambium is deep-seated from the first and the early cork layers push off the original cortex. When this occurs the cambium produces new cells to the inside, and these form the secondary cortex or **phelloderm**. Such a condition is found in species of *Ribes*, the Currants, etc. In

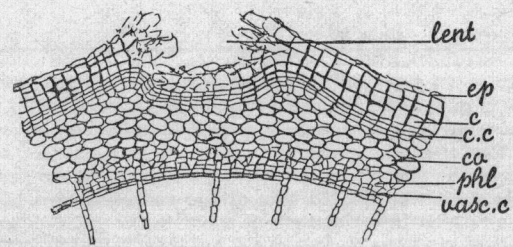

FIG. 26.—Lenticel of Elderberry.

c. cork, *c.c.* cork cambium, *co.* cortex, *ep.* epidermis, *lent.* lenticel, *phl.* phloem, *vasc.c.* vascular cambium.

passing it may be said that the common term 'bark' is variously interpreted and is often used to include all the tissues down to the vascular cambium.

Even in the young twigs the cork is not uninterrupted. In the first year stem stomata are present in the epidermis. Naturally when expansion starts these are weak places in the surface, and below the stomata the cork cambium dips inwards and produces (at any rate while the stem is expanding) a loose mass of powdery cork through which air can pass. This is a lenticel (Fig. 26), and becomes a channel for gas exchange for the tissues of the twig. When radial expansion stops during the summer the pressure of the cork layer is gradually eased, and as the cork cambium still continues to form new cork, the lenticel becomes closed by regular cork layers. It remains in this condition during the winter, but with the rapid onset of radial expansion following the commencement of activity by the vascular cambium in the spring the lenticel cork layers are again ruptured, and we get the same procedure as before. In a smooth bark the position of the lenticels can be followed for years by the great scar-like marks on the outer bark.

Most trees produce some cork, but it varies greatly in depth and texture. Sometimes it develops in a peculiar way, and wings of cork can be seen on the trunk of the Field Maple, and even more fantastic arrangements can be found on some tropical trees. One of the most fascinating forms of cork is found on the Wellingtonia (*Sequoia*), a large conifer which has a bark so soft that it can be punched without discomfort!

The cork of commerce is obtained from the Cork Oak (*Quercus suber*), which grows in the Mediterranean region. The tree produces a thick layer of cork normally, but the original layer is stripped from the trees in the cork plantations when they are about ten to fifteen years old, and after that an even layer of cork is formed and harvested every five or seven years. The sheets of cork are carefully peeled off and then boiled and chemically treated. Ordinary corks are cut parallel to the surface of the sheet (i.e. the surface of the tree), so that the lenticels run from side to side of the cork. Thus when the latter is pushed into the neck of a bottle the glass sides close the lenticels and the cork is fairly rigidly gas- and water-tight. Because the cork layer is of limited thickness a true ' cork ' cannot have more than a certain diameter. Bungs or shives are stamped out from the flat surface of the sheet of cork so that the lenticels run up and down the bung, which is therefore neither gas- nor water-tight.

Secondary Development in Monocotyledons

The vast majority of Monocotyledons do not develop a true cambium; indeed, most of the plants in this group with which we are familiar are herbaceous and do not attain any considerable size. But there are Monocotyledon trees, as for example the Palms, and this condition is usually achieved by a peculiar type of growth which involves the production of new bundles behind a large apical meristem and the subsequent lignification of much of the supporting tissue. It will be observed that these trees are very slender and do not show a marked difference in thickness as they get older. Moreover, the wood is very fibrous and is quite unsuitable as timber. One Monocotyledon tree in particular—Dracaena, the Dragon's-Blood tree—does produce a cambium, but again this gives rise to new bundles rather than to rings of phloem and xylem.

In the bamboos much of the size is due to rapid growth in one growing season, with the subsequent lignification of much fibrous material.

(v) The Internal Structure of the Root

If we start the examination of root structure as we did that of the stem, by taking a longitudinal section of the root-tip, it is found that the arrangement looks much simpler. The root has no lateral developments at its apex, but the latter is covered by a cap of simple parenchymatous tissue which is constantly being worn away by contact with the soil and is continually being renewed by a meristematic layer. The apical meristem resembles that of the stem in general, but is more compact. As before, the tissues quickly show differentiation into dermatogen, periblem and plerome, but on the outside there is another layer, the **calyptrogen,** and this gives rise to the cells of the root-cap.

The differentiating tissues show the same sequence of development as in the stem, but certain differences are soon apparent. The surface layer is now called the **piliferous** layer, and from it the very important root-hairs are formed. These are unicellular outgrowths and are the absorbing structures of the root. Each has a thin lining of protoplasm and a large vacuole or vacuoles. They persist for only a very short time because contact with the soil damages them, and in order that the plant can continue to absorb it is essential that new root-hairs continue to be produced, which means further apical growth. With the loss of the root-hairs the superficial layer of the root disappears.

The cortex is well-defined and is at first entirely parenchymatous, though in an older root the exodermis and hypodermis may become lignified. With the loss of the piliferous layer the exodermis becomes the superficial layer. Perhaps the most striking layer in the cortex of the root is the endodermis. In the vast majority of roots this has developed special features, a condition which is true in other groups besides the Flowering Plants. The endodermis forms a cylinder of tightly packed cells roughly brick-like in shape. Their main characteristic, however, lies in the fact that on the radial and transverse walls a strip of fatty material (suberin) is formed which prevents movement of water along the wall. The general effect of this is that water entering the stele from the cortex must pass through the tangential walls of the endodermal cells and so through the protoplasm. This means that the endodermis can control the passage of the water, and this matter will be dealt with later. In the cortex in general the water can leak easily along the cellulose walls. As the root grows older the endodermis may become so suberised that little if any water can pass across it.

In many roots, especially in the young condition, the cortical cells are full of starch, and in a transverse section the grains can be seen packing the cells. The cells do not contain chloroplasts under normal conditions, though it is not unknown for exposed roots to develop them.

Within the endodermis the stele differs very markedly from the condition found in the stem.

The pericycle is present immediately internal to the endodermis, and plays a considerable part in the later development of the root because of its meristematic activities.

As in the stem, the first vascular elements are produced by the differentiation of a cylinder of procambium extending from the apical meristem. The conducting elements so formed, however, do not appear in bundles, like those of the stem. The phloem and xylem differentiate as separate groups which alternate so that there is a radial arrangement. The number of such groups varies from two to about seven in Dicotyledons, but there may be a large number in Monocotyledons. Each phloem group has a few narrow sieve tubes with their companion cells and are like those of the stem. The xylem arrangement is quite different from that seen in the stem. The protoxylem which develops first is external, just within the pericycle, and subsequent differentiation is inwards, so that the last elements to become mature are those in the middle or near the middle of the stele. If differentiation goes right to the middle of the stele there is a solid star of xylem as seen in the transverse appearance, but some roots do have a pith. The central elements are the largest, and when these have matured the system is, so to speak, closed. The central elements are metaxylem.

All the meristematic tissues have now differentiated and no cambium has appeared up to this time. Between the rays of xylem and the phloem there is merely a layer of conjunctive parenchyma. Thus the whole of the tissue in the root at this stage is primary, and a little examination will show that there can be no obvious continuity with any secondary tissue that may arise.

The actual elongating region of the root is very short, so that little protoxylem is found even in the primary condition.

As the root hairs which are responsible for absorption are very short lived, new ones must be produced as long as the plant is growing at all, and this means continuous growth of the root. It is interesting to watch the enormous root-hair region which often develops in the roots of floating water-plants.

The final differentiation of the primary stele occurs roughly about the region where the root-hairs are lost. It is in this region also that one can see the thickening developing on the walls of the exodermis and hypodermis. As the root has by this time ceased elongation, any secondary tissue will be wholly

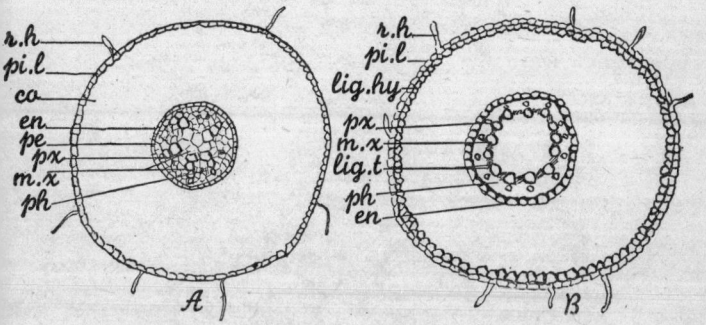

FIG. 27.—Transverse Sections of Roots. Diagrammatic.

A. Dicotyledon, Buttercup. B. Monocotyledon, Maize.

co. cortex, *en.* endodermis, *lig.hy.* lignified hypodermis, *lig.t.* lignified tissue, *m.x.* metaxylem, *pe.* pericycle, *ph.* phloem, *pi.l.* piliferous layer, *px.* protoxylem, *r.h.* root hair.

metaxylem or metaphloem. A transverse section of a young Buttercup root is seen in Figs. 27A and 28.

Primary root-structure is remarkably consistent in the distribution of tissue, so that a root is very easily recognised anatomically. In Monocotyledons the endodermis is much more conspicuous, and the number of vascular groups is usually more than ten. Rarely does the vascular tissue fill the stele, so that many Monocotyledon roots have a pith. The individual groups of xylem and phloem are often very small—the phloem may consist only of a single sieve tube with its companion cell—whilst the xylem may have a small proto-element and a large metaxylem vessel. In aquatic Mono-cotyledons there are few groups, and the roots of aquatic

Monocotyledons and Dicotyledons are very similar.   Fig. 27B
is a diagram of the root of Maize.

The further development of the root

As the primary meristem in the root does not persist, a new
meristem must appear in order that further tissues can be
produced.   This meristem appears in the conjunctive tissue
between the phloem and xylem, and is first seen as divisions

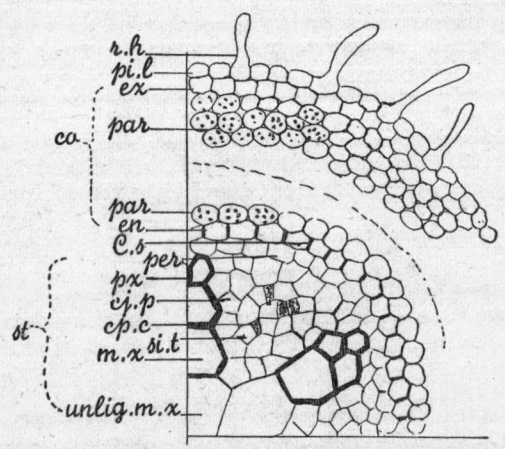

FIG. 28.—Detail of Section of Root of Buttercup.

co. cortex, *c.s.* Casparian strip, *cj.p.* conjunctive parenchyma,
cp.c. companion cell, *en.* endodermis, *ex.* exodermis, *mx.* metaxylem,
par. parenchyma, *per.* pericycle, *pi.l.* piliferous layer, *p.x.* protoxylem, *r.h.* root hair, *si.t.* sieve-tube, *st.* stele, *unlig.mx.* unlignified
metaxylem.

parallel to the surface of the primary xylem.   At first the new
cambium produces elements only towards the existing xylem,
and these differentiate mainly as tracheids, and the cambium is
gradually pushed outwards, so that the primary phloem is
crushed against the pericycle.   Eventually the cambium
becomes a ring enclosing the original primary xylem and the
new secondary xylem.   Further cambial activity now proceeds as in the stem, except that opposite each primary xylem
group a medullary ray is formed by the divisions of the
cambium.   The primary phloem is crushed by the formation

of the secondary phloem and appears as a small group of tissue just within the pericycle.

Thus the progress of secondary development in the root is established, but in most herbaceous plants it does not go very far, and in many cases there is a tendency for new adventitious roots to be added instead of continued development of existing ones. This is not the case, however, in those plants with large tap roots like Dandelion.

A cylinder of vascular tissue comparable with the stem condition gradually appears in woody roots, but the wood does not show the clear arrangement of annual rings that is seen in the

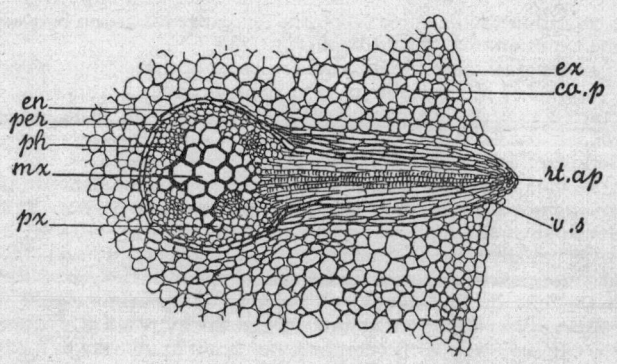

FIG. 29.—Section of Older Buttercup Root showing Origin of Lateral Root.

co.p. cortical parenchyma, *en.* endodermis, *ex.* exodermis, *mx.* metaxylem, *ph.* phloem, *per.* pericycle, *px.* protoxylem, *rt.ap.* lateral root apex, *v.s.* vascular strand.

stem. The wood is much more diffuse in all cases, and frequently much of the tissue remains unlignified, especially in the first year. Presumably there is always much more water available to expand the elements than is the case in the stem. During this cambial development a further activity starts in the pericycle opposite the primary xylem group. The cells begin to divide and a new apical meristem is formed which pushes through the endodermis and across the cortex to become a lateral root (Fig. 29). The vascular tissue becomes continuous with the new elements in the main system, and it is quite usual to see a row of lateral rootlets corresponding to each primary xylem group in the parent root. Such a method of origin of a lateral member is referred to as endogenous, whilst the superficial development of leaves and buds in the stem is exogenous. The origin of lateral roots in Monocotyledons takes place in a similar way, but there is no cambial activity.

As the root grows older it may develop cork, as does the stem, and in many cases an old tree-root may come to look much like a branch. The cork is formed by the divisions of a cambium which originates in the pericycle. Thus as the cork formation takes place the whole of the original cortex is isolated from the internal tissues and is eventually shed. Therefore in an old root the whole of the cortex is secondary, having been formed from the pericyclic cork cambium. Such a condition can be detected by the regular relationship between the cambium and the cortical cells.

In general, then, the internal structure of the root has a very definite and recognisable plan in the primary condition, but is much less easily distinguished in the older root. The tissues themselves have the same structure as those of the stem and no detailed description is necessary.

It should be mentioned that secondary development is rare in Monocotyledons, and the usual change seen is the heavy development of lignified parenchyma and fibres.

One further point remains to be explained. It will have become clear that the vascular arrangement in the root differs from that of the stem, yet the two are continuous. This continuity is easily understood in the woody condition, where the cambial cylinder is complete, but is not at all easy to follow in the herbaceous condition. The changeover takes place in the region known as the **hypocotyl** and involves an inversion of the relative positions of the proto and metaxylem. A series of sections in this region would show how this occurs, but it must be remembered that the condition is static in any one part and is completely masked when secondary thickening has taken place.

In the Monocotyledon the position is not the same, since the stem is not continuous with the main root but ends rather abruptly, whilst a mass of fibrous adventitious roots occur. The vascular tissues of these roots become continuous with the individual bundles of the stem.

(vi) The Internal Structure of the Leaf

The leaf differs from both stem and root in that it is not an axial organ. It arises laterally from the stem and it does not possess an apical meristem. In a great majority of cases the leaf is shed at the end of one season's activity, and is replaced the following year by structures which appear at the new stem apices. In some plants evergreen leaves are found, but they take little part in any further growth which may occur in the stem.

The part of the leaf which remains longest as an active growing region is the extreme base, and in this region the xylem is always protoxylem.

As the young leaves expand it is found that the tissues undergo differentiation similar to that in the stem or root, but, on the whole, differentiation seems to take place from the base to the tip of the leaf. Thus from the leaf-base there is differentiation of vascular tissue downwards into the stem and upwards into the leaf itself. General production of tissues ceases when the leaf stops growing in size, but some differentiation does continue, so that a certain amount of metaxylem is found in the larger bundles of the leaf.

It will be realised that in general the leaf is a flattened structure, thus providing a surface on to which sunlight can fall. The main function of the green leaf is the synthesis of organic material, primarily carbohydrates, and to do this the energy derived from sunlight is necessary. Hence a flattened structure, preferably at right angles to the rays of light, gives a greater opportunity of utilising the sun's rays. Two main types of leaf structure incorporating this principle are found. In one case the leaf has a definite upper and lower surface, the upper surface being the one presented to the light. This type of leaf is said to be **dorsiventral** and its position is typically though not exclusively horizontal (Sunflower, Wallflower, Oak, Beech, etc.). The other type is vertical, and both surfaces are illuminated equally. These leaves are usually elongated and are seen in Grasses, Daffodil, Bluebell, etc. They are described as isobilateral, and the two surfaces are similar in structure.

Other forms of leaf do occur, but are usually associated with special environments.

The structure of the leaf can be studied in several ways. The usual method is to cut vertical sections of the blade and transverse sections of the petiole. It is not always necessary to cut sections across the whole leaf because the arrangement is essentially similar throughout.

Further information, however, can be obtained by stripping off the epidermis of the leaf, a procedure which is particularly useful in studying the openings known as stomata, which will be described later. It is possible to make horizontal sections of the leaf (i.e. parallel to the surface), a useful method of examining vein endings, and also to carry out partial maceration.

If a transverse section of the leaf-stalk or petiole is made, we find two general types of arrangement. In leaves like those of Oak, Ash, Sycamore, etc., there is usually a ring of vascular tissue somewhat similar to the stem arrangement but usually with a straight bundle along the upper (inner or ventral) surface of the petiole. The xylem is directed towards the middle of the petiole. Outside this vascular ring is a cortical layer of parenchyma or collenchyma extending to the epidermis. In the late summer some of this tissue may become lignified. In many other leaves (e.g. Privet, Rhododendron, Laurel, etc.) the vascular system is represented by a crescentic bundle or group of bundles placed in the middle of the petiole and with the xylem facing the upper surface. The cortical region is much wider and is rarely lignified. It may be added that this arrangement is always present in evergreen leaves. Some cambium may be present.

The petiolar arrangement is continued along the main veins, which are usually indicated by ribs of tissue along the back (dorsal surface) of the leaf, but as the vascular strands become smaller they are enclosed within the mesophyll of the leaf. The vascular elements are similar to those found in other organs, but as the bundles decrease in size the xylem elements are found to be tracheids only, and the phloem has usually disappeared before the end of a veinlet is reached.

In the majority of Dicotyledons the veinlets end blindly in the general tissues of the leaf, but in many Monocotyledons the vascular strands are continuous through the finer veinlets and they anastomose to give a closed system.

It must be emphasised that vascular tissues are always enclosed by a tight sheath of parenchyma (in many larger veins a definite endodermis is present), so that even if the strand ends as a single tracheid it is never exposed to an air space. The effect of this is to control the loss of water by the leaf because, as we shall see later, the loss of water is an ever present possibility in plant physiology.

Examination of a section across the lamina of a dorsiventral leaf shows that the vascular tissue is not arranged haphazardly throughout the leaf, but lies at a definite level in the general tissues, which are known as mesophyll. One of the interesting features of the leaf-structure is that much of the vascular tissue which differentiates in the leaf-petiole does not continue into the stem. The amount of vascular tissue passing through the leaf-base is relatively small and the xylem is always protoxylem.

The upper surface of the leaf is covered by an epidermis of cells which fit closely together without air spaces and are often of distinctive shape. Normally these cells are without chlorophyll. A layer of fatty material is secreted and forms the cuticle which cuts down water loss from the surface. The thickness of the cuticle varies considerably with the species of plant and also with the environment. Thus it is found that shade plants have a thin cuticle, whilst leaves exposed to much evaporation have a thick one.

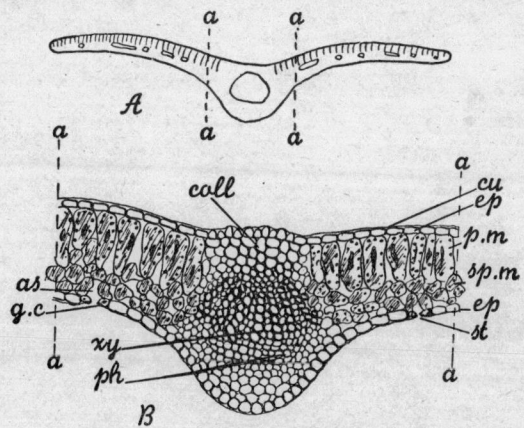

FIG. 30.—Vertical Section of Leaf of Privet.

A. Diagrammatic. B. Detail of Region *aa* ... *aa*.

as. air space, *cu*. cuticle, *ep*. epidermis, *g.c*. guard cell, *coll*. collenchyma, *ph*. phloem, *p.m*. palisade mesophyll, *sp.m*. spongy mesophyll, *st*. stoma, *xy*. xylem.

In the majority of dorsiventral leaves this epidermis is unbroken by the special pores called stomata, which will shortly be described, but it must be emphasised that wide variation exists in this condition and some ordinary land plants such as Pelargonium have a number of stomata on the upper surface.

In the floating leaves of aquatic plants all the stomata occur in the upper epidermis because it is the only one in contact with the atmosphere.

Below the epidermis lies an extensive layer of cells which are typically parenchymatous in structure and arranged in most leaves with their long axis at right angles to the surface. It is true that they are sometimes more cubical but characteristically they are elongated cells, and this has led to the name of **palisade** cells. They are roughly cylindrical and are separated by air spaces, a matter of great importance in the activities of the leaf.

Several layers of these cells may occur, and in the shade of woodlands a leaf may have only a single layer of palisade cells,

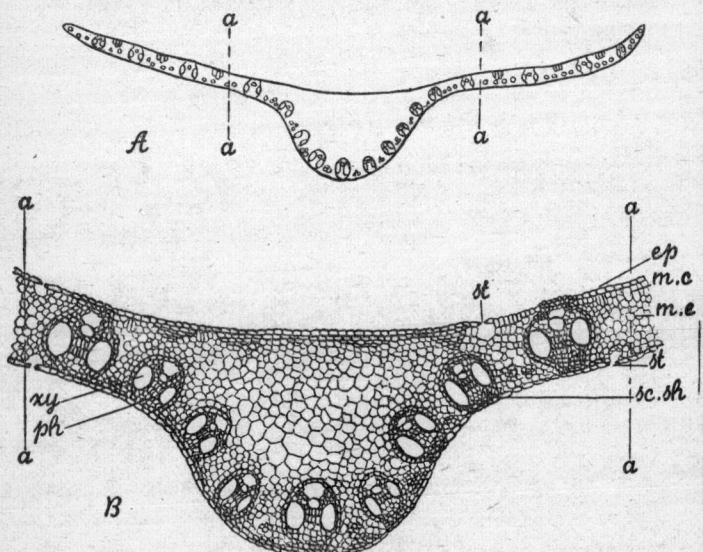

FIG. 31. Vertical Section of Maize Leaf.

A. Diagrammatic.   B. Detail of Region *aa . . . aa*.

ep. epidermis, *m.e.* mesophyll, *m.c.* motor cells, *ph.* phloem, *st.* stoma, *sc. sh.* sclerenchyma sheath, *xy.* xylem.

whilst a leaf of the same species exposed to the sun may have two or three layers.

The most obvious feature of the palisade cells is the presence of **chloroplasts**. These are small protoplasmic bodies more or less discoid in shape which contain the complex of pigments referred to as **chlorophyll**. More will be said about the composition of chlorophyll later, but it is sufficient to say now that it is the pigment responsible for the fixation of sunlight during the process of photosynthesis. As this is perhaps the most notable attribute of plant physiology, the importance of the chloroplasts cannot be over-emphasised.

They are not fixed in position, but can move about in the cell, probably as a result of circulation of the cytoplasm. This movement is related to the angle and strength of the incident sunlight, so that the chloroplasts may be grouped at the inner end of the cell or ranged along the vertical walls. They tend to move, so that they are not directly exposed to the full strength of bright sunlight.

The palisade mesophyll is not continuous over the larger vascular bundles, but does extend over the smaller veinlets.

Below this palisade mesophyll the tissue of the leaf is much less compact, and for that reason is known as spongy mesophyll. The individual cells are parenchymatous and very irregular in shape, sometimes being almost stellate. Large air spaces are present among these cells, so that gaseous circulation in the leaf is facilitated. This mesophyll extends to the lower epidermis, and the smaller vascular bundles lie roughly at the junction of spongy and palisade mesophylls. The cells of the spongy mesophyll possess chloroplasts, but these are fewer towards the lower epidermis, where the light is weaker. In some leaves the palisade is almost indistinguishable in form from the spongy mesophyll, though its cells may be more closely packed.

Around the smaller veinlets the mesophyll forms a sheath, but these cells do not contain chloroplasts. It may be noted that in many aquatic leaves, especially floating leaves, the air spaces are even larger and the spongy mesophyll is arranged in bars across the spaces, with a denser layer around the vascular strands.

Finally there is the lower epidermis. The majority of its cells are similar to those of the upper epidermis and are without chloroplasts, but at frequent intervals, though varying in density in different species, are the stomatal openings. A typical stomate is shown in Fig. 32A. From the surface view the pore is enclosed by two somewhat crescentic cells called guard cells, and these possess chloroplasts. The walls of these guard cells are unequally thickened so that the wall adjacent to the pore is much strengthened, whilst that distant from the opening is less thick and the cell can dilate in that direction.

In section the stoma is seen to be somewhat sunk below the general level of the epidermal cells, and the outer surface of the guard cells is again thickened and protected by cuticle which overhangs the opening (Fig. 32B).

The form of stomatal structure varies somewhat, but the

principle is the same throughout. Frequently the epidermal cells immediately associated with the guard cells have a distinct arrangement, and this is used in the identification of certain leaves or parts of leaves used as drugs.

The stomata control the movement of the gases into and out of the leaf, and the size of the opening itself is, of course, determined by the change of shape of the guard cells. This is due to changes in the water content of the guard cells, and this matter will be discussed in the chapter on plant physiology.

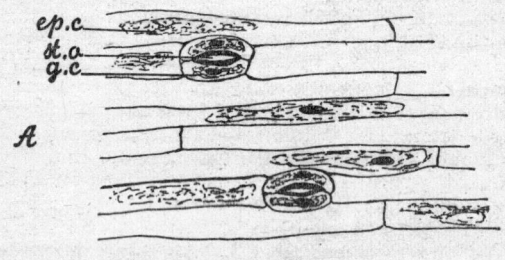

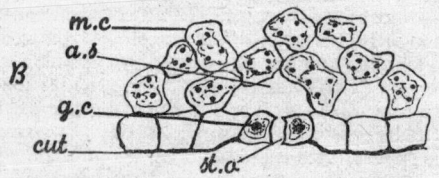

Fig. 32.—Stomata from a Monocotyledon Leaf.

A. Surface view. B. Section of Epidermis, etc.

a.s. air space, *cut.* cuticle, *ep.c.* epidermal cell, *g.c.* guard cell, *m.c.* mesophyll cell, *st.o.* stomatal opening.

The stomata may be protected by hairs on the leaf surface, by being restricted to grooves in the leaf structure or by being sunk well below the epidermal level.

There is always a large air space behind the stoma, and this is continuous with the other air spaces in the leaf. The structure of the stomata is similar in all parts of the plant in which they occur—leaves, young stems, flower petals, etc.

As the leaves grow older there is a tendency for more ligni-fication to occur, particularly in the neighbourhood of the bundle, but in the very base of the leaf there is no additional lignification and the tissues remain more or less juvenile and the xylem is always protoxylem. In those leaves which are shed annually (deciduous), such as Oak, Ash, Sycamore and many others, a sheet of small cells can be distinguished across the base of the leaf. This is called the absciss layer and marks the position at which the tissues will break down and lead to the fall of the leaf. The mechanism of leaf-fall is rather complex, but one feature is the obstruction of the vascular strands towards the end of the life of the leaf accompanied by a general tendency of the leaf to become very dry. In herbaceous plants the leaves generally wither on the plant without any definite shedding whilst in evergreen plants the process is very slow, though in the end the procedure is much the same and is assisted by the strain on the petiole due to expansion of the stem.

When the leaf falls a characteristic scar is produced, and eventually this is healed over by a deposit of cork continuous with the cork on the twig.

So far the discussion of leaf-structure has dealt mainly with the dorsiventral leaf and the modification which may occur in special circumstances. Thus in floating aquatic leaves the stomata are on the upper surface immediately adjacent to the palisade tissue and the vascular strands are frequently enclosed in a well-defined endodermis. In submerged leaves the mesophyll is undifferentiated, the leaves are very thin or finely divided and stomata are absent.

In plants growing in salt marshes or on sand dunes the leaves are often swollen, due to the presence of parenchyma cells full of water, there are few stomata and the palisade is poorly differentiated. In many cacti and similar plants the leaves may be reduced altogether, and it is then found that the stems take on the function of the leaves without having the extensive (and as far as water loss is concerned, expensive) surface area.

In other cases the leaves roll during periods of drought (many Grasses), whilst the Heaths have leaves which are always inrolled. In such cases the stomata are on the enclosed surface, so that again water loss is reduced.

The **isobilateral** leaf differs in the distribution of the tissues in the leaf.

In the first place, stomata are present on both surfaces and the palisade is poorly differentiated from the rest of the mesophyll, but is represented on both surfaces of the leaf. Moreover, although the large bundles may be centrally placed, smaller bundles may be present in two rows, a row under each surface. It should be noted, however, that the larger bundles are orientated so that the xylem faces the inner or ventral surface of the leaf as in the dorsiventral type, but where there are smaller bundles the xylem is towards the inside of the leaf.

In the Grasses the stomata have guard cells of a distinctive dumb-bell pattern.

In some isobilateral leaves there is a superficial resemblance to the dorsiventral type. Thus the leaf of Maize has a distinct rib or vein along the " back " of the leaf, and in this region the bundles lie along the lower surface. In the lamina the leaf is typically isobilateral.

The structure of the isobilateral leaf means that both surfaces are equally capable of synthesis, and this is undoubtedly associated with the upright position of the leaf. Most of the isobilateral leaves are found in Monocotyledons, though not all the latter have leaves of this type. A few peculiar forms do occur, as for example the cylindrical or centric leaves of Onion and the Rushes where there is only one surface. The anatomical derivation of some of these Monocotyledon leaves is a matter of argument. There is little doubt that structually they cannot be resolved into the same organisation as the typical Dicotyledon leaf, although they perform the same functions.

Vertical sections of the leaf of Privet (dorsiventral) and Maize (isobilateral) are shown in Figs. 30 and 31.

With the above we conclude a brief review of the general anatomy of the vegetative structures of the Flowering Plant and the different types of cell formation and organisation which are to be found. It cannot be over-emphasised that wide variations are to be found in different plants, especially in relation to environmental differences, but these are matters which can be followed up in more detailed anatomical works, to some of which reference is made at the end of this book.

It must also be remembered that there is a complete continuity between the various organs, and that the structural distribution of vascular tissues in one organ is such that they are in communication with those in all other parts of the plant.

In the succeeding chapters an attempt will be made to show how the functions of the plant are associated with the structures which have been described. It will be found that both living and dead tissues play a part in the life-processes of the plant, though it must be realised that the existence of a particular organ does not imply that it must have a function—or at any rate that the function is known. It might even be said that much time is wasted in trying to ascribe functions to every structure developed by the plant, particularly as even now we do not know all the factors which govern the development of plant tissues and organs.

3. PHYSIOLOGY—HOW THE PLANT LIVES

No matter how much one can learn by studying the tissues and the general structure of the plant, this gives only a limited picture of the plant's method of growth, of how it obtains its food materials and produces energy and, indeed, of all the processes which constitute life. Some of these activities are very complex, so much so that many of the chemical changes cannot be reproduced in the laboratory. Thus some of the syntheses which take place every day in the plant and are of the utmost importance to man are still beyond his knowledge.

One of the first things that must be realised about the vast majority of plants is that they utilise only the most simple substances in the building of the various materials which are found in their tissues. Most of these substances are taken in by the roots from the dilute solutions which are present in the soil, though some enter the plant as gases through the surfaces of the aerial parts.

Before discussing what happens in the plant, it is necessary to consider the environment in which it lives. Though mention has been made already of the variations which may be present in that environment, there is a lot of basic similarity in the principles involved.

(i) Soils

In the first place, the plant grows typically with its roots in the soil and with its stem and leaves in the air. The air is a mixture of gases, the principal constituents being nitrogen (about 79%), oxygen (about 21%), carbon dioxide (about 0.03%), with traces of other gases. These proportions vary under different conditions but not very greatly—the greatest difference probably being in the amount of carbon dioxide which is produced by the respiration (" breathing ") of animals and plants and by the breakdown of carbon substances, particularly in the soil. Thus in regions of dense vegetation there is often more carbon dioxide in the air than in sparsely populated areas, the carbon dioxide coming in part from the living plant and in part from decaying plant remains. On the other hand, the soil conditions can vary enormously. Soil is a mixture of mineral matter produced from the weathering of rocks and organic matter derived from the breakdown of plants and animals. This latter soil constituent is called **humus,** and varies from recognisable remains to a fine debris from which certain chemical substances ultimately pass into the soil. The mineral fraction again is very variable. The coarsest fraction is called gravel, the next finer constituent is called sand, then there is silt and finally the smallest particles form the clay. Each particle has normally a fine film of moisture around it, and in this film various substances form a very dilute solution. It is from this solution that the roots of the plant absorb the materials required for its life-processes.

The composition of the soil is of great importance. Its texture depends on the proportion of the various fractions present and on the chemical nature of the rock from which it is derived. This may or may not be the actual underlying rock. Many areas have soil which has been carried down by rivers from other regions, and there is then no connection between the top soil and the rock far beneath. Such deposits are known as alluvium and are usually very fertile.

If the soil is coarse (i.e. having large particles), then it is well aerated, but retains little water, e.g. sandy soils and gravels. As the proportion of small particles increases the soil holds more water, but aeration becomes more difficult, and at the other end of the scale there is the clay with particles so fine that they will remain suspended almost indefinitely when a sample is shaken up with water. Clayey soils easily become waterlogged and tend to be cold and airless, with the result that they are inclined to be infertile. It is possible to influence the growth of the plant by varying the proportion of the mineral fractions and also the humus. The addition of manure, animal and plant material to the soil not only provides a source of chemical substances, but also affects the physical structure of the soil. If a sample of the soil is vigorously shaken up in a narrow vessel and allowed to stand, the fractions will settle out with the coarsest particles at the bottom and the finest at the top, the colloidal clay particles remaining suspended for a very long period. The grosser humus pieces float. A typical blended soil is called a loam.

In the same way the chemical nature of the soil has a profound effect on plant growth, and a great difference will be found between the plants which flourish on basic soils (represented in our country mainly by chalk and limestone areas) and those which appear on acid soils which are represented typically by moorland areas. In another direction we get very saline regions—e.g. salt marshes—and again there is a specialised plant population. These represent extremes, and within this compass there is a wide range of conditions, often very much modified by the activities of man. The study of these conditions is known as ecology, and a further reference will be made to the types of habitat.

What are the substances which the plant requires and how can we determine them?

In the first place, we can analyse the plant tissue by chemical methods and find which chemical substances are present. By such means we find that carbon (C), hydrogen (H), oxygen (O), nitrogen (N), sulphur (S), phosphorus (P) are always present and are actually part of the living structure. Other elements are essential and often present in some part of the structure. Thus magnesium (Mg) is present in the green pigment chlorophyll, and iron (Fe), though not part of this substance, is essential for its formation. Sodium (Na), potassium (K), calcium (Ca), chlorine (Cl) are all found, though not necessarily in the structure. In addition, many elements are required in very small quantities—the so-called trace elements—and some of these which are beneficial in small quantities may be poisonous if present in concentrations of more than a few parts per million.

These substances are present in the soil as simple salts in solution, and among the most frequent are chlorides, carbonates, sulphates, nitrates and phosphates derived by a multiplicity of chemical actions in the soil and found normally only in very dilute solution. Carbon is probably taken in principally as carbon dioxide from the air. In the plant cells there is also a solution—the cell sap—and here there is a similar range of dissolved materials (though often in surprisingly different proportions), together with other substances such as sugars and proteins and their derivatives which have been synthesised by the plant.

The requirements of the plant can be determined by carefully controlled experiments in which plants are grown under conditions where a particular element can be withheld. This procedure is not an easy one to carry out on a small scale, and it must be admitted that school laboratory experiments are often inconclusive.

The experimental procedure consists of growing batches of seedlings or older plants similar in size, etc., in a series of vessels which contain culture solutions or sterile sand watered by the appropriate solutions.

A control is supplied with a solution containing all the necessary elements, whilst the remaining plants are kept short of one element at a time by judicious balancing of the salts used. The experiment usually takes some time, but when carefully carried out the results are quite striking. Thus in the absence of iron chlorophyll does not develop and the plants are yellowish, in the absence of nitrogen growth is stunted due to the inability of the plant to produce protoplasm. Again, excess of nitrogen will tend to delay flower production, and each element seems to have some influence. Much has been learned from these methods and indeed it has been found possible to concentrate and to some extent standardise plant growth by growing certain crops (e.g. tomatoes and carnations among others) in balanced culture solutions—a practice known as **hydroponics** and exploited in some degree on a commercial scale.

The proportion of these substances in the soil varies a great deal and is influenced to a considerable degree by their solubility. Thus very soluble salts are easily washed out, and this is a serious problem in the case of nitrates. Although the higher plants can absorb and utilise amino-acids and nitrates, there is little doubt that their principal source of nitrogen is from nitrates. As nitrate is easily leached out of the soil, and as its appearance is dependent on the activity of bacteria, which are also affected by soil conditions, it is obvious that the supply of nitrate is always a problem.

The elements themselves may be held in the chemical complex of the clay particle and are available in varying degrees to the plant.

Changes are always taking place in the soil. Besides the larger plants which grow on the surface of the ground, the soil provides a home for countless microscopic organisms, fungi, bacteria and animal organisms such as protozoa. They are concerned in vast series of changes which result in the breakdown of humus to the simple substances, such as carbon dioxide from carbon materials, ammonia and ammonium compounds from proteins and sulphuretted hydrogen from sulphur compounds. In addition to these changes, some of the mineral salts are affected by bacterial action. Some of the substances formed as a result of bacterial action are then used by other bacteria and soil organisms as sources of energy, so that further changes occur, leading to the formation, for instance, of nitrites and then nitrates from ammonium compounds. These changes depend on soil conditions; thus in the case of nitrate formation the organisms concerned require a well-aerated soil containing plenty of humus. This change is generally referred to as nitrification. If the soil is badly aerated and acid—a condition which may well arise because it is waterlogged—other organisms develop which obtain their energy at the expense of the nitrate, and this may be broken down to nitrogen itself and lost altogether. This is called denitrification. Yet again, there are bacteria which can bring free nitrogen (from the soil-water) into combination, a process which is called nitrogen fixation. The classical example of this is an organism which is called *Bacillus radicicola*, which is found in nodules on the roots of plants of the Bean family (Leguminosae). Presumably there is an exchange of materials between the micro-organism and the host root, the usual supposition being that the former obtains carbohydrate synthesised by the green plant, whilst the latter absorbs nitrogenous materials formed in the bacterial cells. This relationship plays an important part in the crop economics of the soil.

In short, there are what are called cycles of chemical change, of which part occur in the soil and part in the plant (and animal). They can be illustrated by diagrams as shown.

The carbon cycle can be depicted as follows:

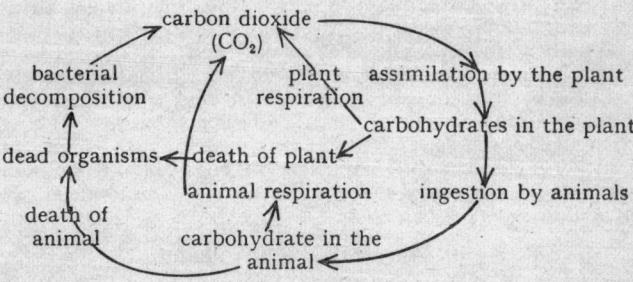

It should be pointed out that some bacteria are capable of chemosynthesis of carbon compounds which will be broken down by bacterial respiration or by death of the bacteria and subsequent decomposition as above. These chemosynthetic bacteria present a separate problem in the study of biological metabolism.

The Nitrogen Cycle

This is much more complex than the carbon cycle because of the many " short cuts" which may occur. The starting point, or perhaps one ought to say the simplest stage in the cycle, is ammonia or ammonium compounds.

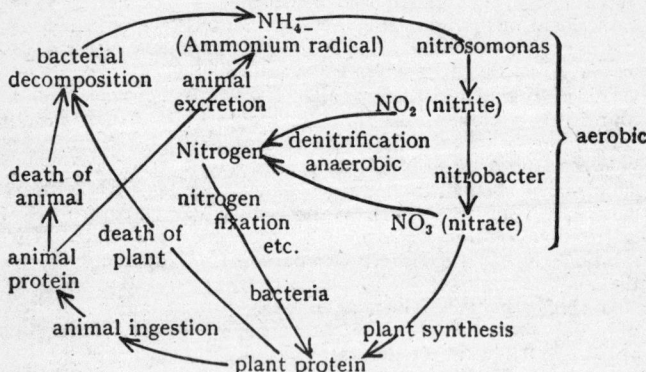

This represents the main aspects of the cycle, though there are many minor deviations.

Finally it may be said that in acid and waterlogged soils which become anaerobic decomposition may be held up (partly because bacteria do not grow well in acid conditions) and extensive preservation of plant tissues may occur. This is particularly well seen in peat bogs and finally, of course, in coal. The presence of marsh gas in bogs and firedamp in mines (both are names for methane, CH_4) is due to anaerobic breakdown.

(ii) Absorption by the Plant

It has been pointed out that the greater part of the material taken in by the plant is absorbed by the root-hairs, which cover only a very short region just behind the growing part (or, more correctly, the elongating part) of the root. These root-hairs are in contact with the soil water, the dilute solution which covers the soil particles and from which the root absorbs its water and salts.

The method by which this is achieved depends on a number of factors, and in particular the peculiar properties of the cytoplasmic surface layer. When substances dissolve in water they diffuse through the solvent to equality of concentration. If a crystal of a coloured salt like potassium permanganate is dropped into water there is at first a dense colouration around the crystal, indicating a high concentration of solution. Gradually the crystal dissolves and the molecules diffuse through the water until there is an equal concentration, as indicated visibly by an even colour throughout. If the water is agitated (e.g. by stirring), the diffusion takes place much more quickly. If a second salt is dissolved in the water, it also will diffuse in the same way and independently of the first one. This property of diffusion plays a very important part in the movement of salts into the plant and from cell to cell, but of course it is not free diffusion—the movement is governed by the cytoplasmic membrane, and the passage of the dissolved material is very complex. Moreover, the movement of water itself is affected by the presence of dissolved substances. If a solution of common salt is enclosed in a membrane like pig's bladder or cellophane and suspended in distilled water, the cell so formed will expand or, if the apparatus shown in Fig. 33 is used, the water will rise in the tube. This movement of the water is called osmosis, and theoretically the water will continue to enter the solution until the latter is as dilute as the distilled water (which of course is really an impossibility). In practice the intake of water is halted because the salt diffuses out-

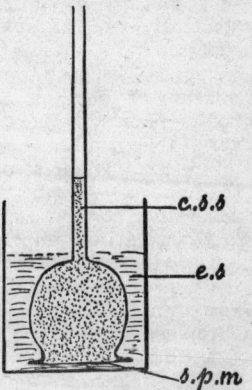

FIG. 33.—Experiment to Demonstrate Osmosis.

c.s.s. cane-sugar solution, *e.s.* external solution, *s.p.m.* semi-permeable membrane.

wards until its concentration is the same on both sides of the membrane. The same process would occur if the distilled water was replaced by a solution more dilute than that inside the membrane, but the rise would be halted more quickly. The difference in the rate of movement between the water and the salt is due to the fact that the membrane is more **permeable** to water (the solvent) than to the salt (the solute). If cane-sugar replaced the salt the rise would continue longer because the membrane is much less permeable to cane-sugar. Such a membrane is said to be **semi-permeable,** being readily permeable to the solvent but less so to the solute. From this simple demonstration a whole range of conditions can be developed. When substances such as those mentioned are dissolved in water they exert what is called an osmotic pressure, which can be measured in terms of the concentration of the dissolved material. Thus the molecular weight in grams of a substance like cane-sugar exerts a pressure of 22·24 atmospheres when dissolved in a litre of water. With a substance like common salt the pressure is greater because in a relatively dilute solution the salt is not present as the molecule NaCl but as Na· ions and Cl⁻ ions, and each exerts its own effect. In a mixture of substances in solution each behaves independently, and the osmotic pressure of the cell is the combined effect of the solutes. This is of the utmost importance in the living cell, because the solution in the cytoplasm is always a mixture of inorganic salts and organic substance synthesised by the plant. In the cell the cytoplasmic surface is the semipermeable membrane, and its relative permeability depends on many circumstances, probably the simplest being the size of the molecule of the solute. Although we speak of semipermeable membranes it is unlikely that living membranes are ever entirely impermeable to substances in true solution, but obviously some move much more slowly than others. It must be emphasised that the cytoplasmic membrane is living and probably variable in its permeability. Certainly when the cytoplasm is dead it is entirely permeable.

Let us now consider the case relative to the living cell. The cytoplasm secretes and is supported by a cell wall, which in general is completely permeable to the various solutions which bathe it. (Exceptions to this general permeability are found in those walls which occur, for instance, at the plant surface, and in which fatty materials are deposited which prevent water loss. In such cases they are impermeable to everything.) Normally, therefore, the cell wall exerts no control over the movement of water and salts. The controlling factor is the cytoplasmic surface, and it must be emphasised that the movement of water is entirely independent of the movement of salts, the water movement being affected by the osmotic pressure and the salts moving according to their individual concentrations.

In the plant there are millions of living cells, each of which contains a solution in the protoplasm called the cell sap. This solution contains mineral salts and organic substances such as sugars and amino-acids which have been synthesised by the plant. Constant exchange of materials takes place between the cells by diffusion, and water moves by osmosis. The concentrations vary all the time, but it is probable that they are higher in leaves and fruits. The ability of the cell to absorb water will depend on the difference between its osmotic pressure and that of systems adjacent to it. But other factors affect the entry of water. As water enters the cells the protoplasm expands and begins to push against the wall, and the latter, being solid, exerts a counter-pressure. As this back pressure increases, water will enter the cell more slowly until a point is reached where the resistance of the cell wall balances the force tending to bring water into the cell. Of course if the cell wall is weak and the osmotic force high it is quite probable that the cell wall will burst. Cells of seaweeds, which necessarily have a high osmotic pressure, will burst when placed in fresh water.

These various factors can be summed up in this way: the actual tendency of the cell to take in water is measured by the difference between the osmotic pressure of the cell sap and the wall pressure. This difference is called the **suction pressure** and the system can be expressed:

Suction pressure (SP) = osmotic pressure (OP) — wall pressure (WP) and when WP equals OP, then obviously SP will be zero.

It will probably have occurred to the reader that in the plant some cells will be losing water to adjacent ones with a higher suction pressure. If cells are exposed to excessive external concentrations, the cytoplasm loses water and eventually shrinks away from the cell wall and collapses. This is called **plasmolysis**, and can cause the death of the cell. When the cytoplasm is expanded against the wall, that is, completely filling the cell, the latter is said to be **turgid**, and this helps to a considerable extent in keeping the plant organs rigid. When water is lost the cells become flaccid and the organ undergoes a condition known as wilting. This phenomenon is a familiar one in leaves exposed to conditions of rapid evaporation.

To return to the root-hair. The dilute soil solution which surrounds it has a much lower osmotic pressure than that of the hair cell itself, and as a result of this water passes into the hair by osmosis. As the root-hair takes in water its cell sap becomes more dilute, and thus it is possible for water to pass to the next cell inward and so across the cortex. As the walls themselves are absorptive, water can leak from one wall to another, as it does in blotting-paper, but as these walls are normally saturated this movement may be significant only in the upward irrigation of the cortex.

Eventually the water moving inwards reaches the endodermis, and here there are complications. It was pointed out in the section on root anatomy that a fatty deposit in the radial and transverse walls allowed the passage of water only through the protoplasm of the endodermal cell. Thus these cells probably act as one large semi-permeable membrane separating the stele from the cortex. It may be that the stele with its vascular system has a higher osmotic pressure than the cortex, so that water tends to move into the stele as it has done across the cortex. It must also be borne in mind that the absorbing circumference of the root is much greater than the stelar surface, so that there may be some additional force tending to push water into the stele. The method is not clear or fully understood, but certainly a pressure develops in the stele sufficient to bring about the differentiation of the xylem elements and at certain times to produce an upward pressure, which is called root pressure. It is generally claimed that the upward movement of water is assisted by the demands of the leaves. Certainly the loss of water from the leaves and the measurement of osmotic pressure in the leaves suggest that it is enough to maintain a column of water sufficient to reach the top of a tree. One of the most difficult things to grasp is the continuity of the whole process. We are not faced with the sudden absorption of water into a ready-made system, like some preformed mechanism, but with the gradual growth of an absorbing system which is developing as it functions. That upward movement of the water does take place in the xylem can be demonstrated quite easily. If a cut shoot bearing leaves is placed in a coloured solution (e.g. dilute eosin) it is possible to see the progress of the solution along the vascular strands, and if sections of the tissues are examined microscopically it can be seen that the solution is confined to the xylem vessels and tracheids. A more important point is that the same effect can be obtained by putting whole seedlings with their roots in such a coloured solution. If the plant be left in the solution, the colour can be traced into the finest xylem elements of the leaves. If a flower such as a Daffodil is used, a very striking effect can be produced.

The movement of water through the plant can be attributed to a number of factors. It should now be clear that there is a number of vascular elements, xylem vessels or tracheids, giving a continuous system through the plant but interrupted even in vessels by cross walls. This is probably not a great drawback, since lignin will allow the passage of some water and the walls will normally be saturated. In tracheids the pits may play a great part in the movement of water. Altogether there are innumerable fine columns of water extending upwards in tall trees for 200–300 feet. At 300 feet such a column of water is equivalent to an osmotic pressure of ten atmospheres. This is easily exceeded in most leaves. Furthermore, Dixon and Joly demonstrated that a column of water could stand a tension of 300 atmospheres before breaking, so that the columns in the tree are not in danger of collapsing. In addition, root pressure helps to some extent. It does seem that the conducting elements are nowhere exposed to direct evaporation, so that leakage is minimised. Nevertheless much of the mechanism of water movement is still not understood—for instance, it is suggested that during active water loss by the leaves the xylem elements may contain only water vapour. It is of the utmost importance that air should not enter the elements, otherwise there would be an airlock, a development which may be less obstructive in a tracheid than a 30-foot vessel, since the bubble does not pass the end wall. More will be said of the loss of water and its effects later.

So much for the movement of water—what is the mechanism in the case of the dissolved substances? It has already been stated that the solutes diffuse independently both as molecules and as ions. Thus if the concentration of any ion in the root-hair is less than its concentration in the soil water, that ion will diffuse into the cell. If it then just stayed in solution there is a possibility that further entry would be blocked, due to an equilibrium having been attained. But several things may happen to the ion (or molecule). It may be used by the cytoplasm in synthesis or otherwise combined, it may be adsorbed on the surface of a colloidal system, or it may even be deposited. In any of these cases it is no longer in solution, and therefore does not affect further inward diffusion. Because of this, cells may accumulate ions to an analytical concentration many times that in the soil-water (or, in the case of seaweeds, in the surrounding sea-water), though the concentration *in the cell sap* may be low. In the plant these substances pass by diffusion from cell to cell and also by translocation in the xylem.

Organic substances are in the main synthesised by various cells in the plant, but there is a good deal of movement. It is probably that actual translocation is carried out in the sieve tubes, and diffusion from cell to cell is probably as glucose in the case of carbohydrates, amino-acids in the case of proteins and fatty acids in the case of fats. It must be remembered, however, that the plant has very considerable abilities in the matter of conversion from one form to another, and, as in the case of salts, the organic substances may be deposited as insoluble materials for storage, even though this may be for only short periods. Thus in most leaves starch appears during sunlight as a temporary product of photosynthesis, probably because its retention as sugar would block the photosynthetic process. At night the starch is converted into sugar and usually translocated to a more permanent storage area.

Summing up, then, it may be said that although all movement of materials depends on solution, any individual solute moves according to its concentration and independently of the water movement—with the exception that there may be " stream " movement in the xylem. The whole system is constantly changing, and the behaviour of each cell is governed by the factors immediately around it.

An important point to remember in root absorption is that it can take place efficiently only in well aerated soils. Oxygen is essential to the living root and it may well be that respiratory energy may be used in carrying out absorption against ordinary diffusion gradients and in the movement of solutes from cell to cell. The passage of solutes across the cell membrane may be affected by various processes requiring energy and constituting what is known as active transport.

(iii) Transpiration and the Loss of Water

If a wet surface or a sheet of water is exposed to the air, the water will evaporate unless the air in contact is already saturated. Many factors can affect the rate of evaporation besides air saturation—such as the temperature, the movement of the air, etc.

Within the plant there is an extensive system of air spaces with which many of the living cells are in contact. Since the cell walls are wet, evaporation will take place into these air spaces if the latter are not saturated with water vapour. This would soon happen if they were not in contact with the open air. We have seen that the leaf possesses many stomata in the surface, and when these are open gas exchange is possible between the plant tissues and the atmosphere. So when the stomata are open and physical conditions favour evaporation the plant will lose water, and this is known as transpiration, a process which has been extensively investigated, particularly with respect to its effects on the plant.

It is quite easy to demonstrate some of the features of transpiration. If a leafy twig is enclosed in a glass bell-jar in such a way that there is no evaporation from the vessel in which it is standing, and the whole apparatus is then exposed to light, drops of water will eventually appear on the surface of the bell-jar. Such an experiment should be put side by side with a control—a similar apparatus with no plant. Other experiments can be devised to show that the loss of water is directly related to the distribution of the stomata. Thus it is possible to make use of the fact that cobalt chloride is pink when wet but blue when quite dry. If strips of dry cobalt chloride paper are carefully attached to the surface of, say, Cherry Laurel leaves and the plant then exposed to light, it will be found that the rate of evaporation as indicated by the appearance of the pink colour is much greater on the under-surface of the leaf, and microscopic examination will show that stomata are restricted to that surface. In the case of an Iris leaf it will be found that transpiration proceeds fairly evenly from both surfaces, and that this is again reflected in the more or less equal distribution of the stomata. Further information can be gained by isolating leaves and smearing the surfaces with some substance such as vaseline which will block the stomata. By treating the lower surface in some cases and the upper in others, it is possible to see which surface loses the most water. In adult leaves it is found that little water is actually lost through the cell walls and cuticle of the leaf surface, but such loss is much greater in young leaves and in leaves normally growing in considerable shade where the cuticle is thin.

There are also methods of finding out the rate of water loss. Several instruments have been devised by which it is possible to estimate the amount of water lost in a given time. These instruments are called potometers, but care must be taken in

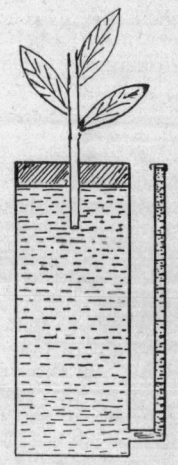

FIG. 34.—A Simple Potometer.

using a potometer to see that the actual water loss is measured, and not merely the water taken in by the plant from the vessel. Fig. 34 shows a type of potometer in which the water loss is determined by weighing the apparatus, whilst the fall in the level of the water in the graduated limb indicates the amount of water taken in by the plant.

Many other experiments are possible, but these indicate some of the major features.

A more difficult problem is to assess the part played by the stomata in relation to water loss. Obviously when the stomata are open and atmospheric conditions favour evaporation, water will be lost by the plant, but investigations have shown that the opening of the stomata is not directly related to the conditions which govern loss of water. Quite frequently the stomata are open when water loss is occurring at a rate which is prejudicial to the health of the plant, and in any case the rate of water loss is not affected by the closing of the stomata until the actual pore size has decreased by more than 50%. On the other hand, on many occasions when the atmosphere is fully saturated and transpiration would be negligible it is found that the stomata are closed.

Thus it appears that the stomata, although mechanically able to control water loss, are not physiologically geared to the factors which affect transpiration. It will be seen later that their opening is actually related to the conditions which govern photosynthesis. One very important point is that the stomata are generally closed at night and open during the day (the latter subject to other conditions), the movement being determined indirectly by the photosynthetic activities in the guard cells.

Most workers agree that on the whole transpiration is a process that may at times adversely affect the plant. On a warm bright day the plant may lose water to such an extent that the leaf-cells become flaccid and the leaf wilts. Moreover, experimental evidence suggests that on such days the vessels and tracheids of the xylem are under tension—they are not full of water. Whether they contain water vapour and fill again at night is open to argument, but it is well known that whilst some shoots will " bleed " when cut in the spring, liquid will enter the shoot if the cut is made later in the year, when the plant is in full leaf, and if the cut is made in daylight. (The actual method is to make the cut under a liquid such as Indian ink held against the wood surface in a plasticine cup.)

The question of whether transpiration fulfils any useful purpose produces varied opinions. It is frequently said that the continued evaporation from the leaf-cells leads to an increase in osmotic pressure of the contents, so that more water is eventually withdrawn from the xylem. Hence it is claimed that a stream of water is maintained up the plant and that transpiration therefore helps in the distribution of the substances in solution. There is evidence that the movement of water in the tree is affected by the osmotic pressure in the leaf-cells, and it has been stated that the higher the leaf on the tree the higher the osmotic pressure of its cells.

But on the whole it is probably true that transpiration is a process which the plant cannot prevent and in some cases its effects may be serious.

It should be noted that many plants show features which tend to reduce transpiration. Such modifications are most commonly found in species which grow in conditions where transpiration is likely to be excessive or where absorption of water is difficult. These modifications are of several kinds, but the general effect is to create a local area of saturation on the stomatal surface so that water loss is reduced. Thus in some cases there is an extensive covering of hairs; in others, such as the leaves of Marram and other grasses, the leaf folds or rolls, so that the stomata are enclosed during dry periods. In many cases the leaf-surface is much reduced, as in most Conifers and in plants like Gorse and Broom, whilst in Bilberry the leaves are shed when water shortage develops. Such plants are said to show **xerophytic** characters.

Besides a too-ready loss of water, the plant may have difficulty in obtaining it. This may arise from more than one cause. In deserts there is an actual shortage of water, and one finds plants with reduced leaf-area and often with water-storage cells. Such plants frequently have an enormously enlarged root system. In salt marshes and on seashores there there is a physiological drought because the water is highly saline, which means that it has a higher osmotic pressure than the root-cells, so that absorption is difficult. Here again the succulent type of plant with reduced leaves is common.

The important factor is that the plant cannot lose water beyond a certain level, otherwise the cytoplasm is damaged, and it would appear that in trees at any rate the first organs to be affected by severe drought would be the leaves which would wilt and then wither. In herbaceous plants prolonged loss of water results in the withering of the whole plant, as can be seen in some of the smaller annuals during a hot dry summer.

(iv) The Synthesis of Organic Material

One of the most important aspects of the life of the green plant is its ability to build up organic substances from simple inorganic sources. It has already been explained that the plant absorbs salts from the soil solution and that carbon dioxide and oxygen enter the tissues from the air (or from solution in some cases) Within the cytoplasm these raw materials are synthesised into the various compounds which are required for the plant structure and to provide energy. The structural requirements include the formation of new protoplasm as well as the non-living parts such as the cellulose wall, lignin, suberin and other deposited materials.

Probably the most specialised stage of the synthesis is the

formation of the carbohydrate, although modern knowledge indicates that the production of the other principal substances, fats and proteins, actually starts before the carbohydrate molecule is completed. In other words, after the initial stages of synthesis the processes diverge towards the respective final products.

The synthesis of carbohydrate is usually referred to as photosynthesis or carbon assimilation, and occurs in all green plants in sunlight. In most plants the completion of this process is indicated by the appearance of starch in the leaf-cells, though a few species never synthesise starch. It is probably preceded by a sugar such as glucose, but this does not accumulate.

It has been generally recognised that the synthesis of carbohydrate is dependent on the presence of light, the green substance chlorophyll, a supply of carbon dioxide and water, together with other factors which probably affect the rate of synthesis but do not actually control it.

A number of comparatively simple experiments can be performed to show how the factors mentioned are concerned in photosynthesis, and for this purpose plants with fairly large and thin leaves, such as Nasturtium (*Tropaeolum*) or *Pelargonium*, are very convenient. It must be remembered that when experiments are carried out to illustrate certain aspects of a process it is necessary to put up control experiments as far as possible in which the specific factor under consideration is not restricted.

In the simple photosynthetic experiments the presence of starch in the leaves is taken as the criterion that carbon assimilation has taken place, and the first necessity is a method of demonstrating the presence of starch. To do this a leaf is taken from the plant and dipped in boiling water to kill it (stopping any further changes) and to make easier the removal of the chlorophyll. The leaf is then transferred to warm 90% ethyl alcohol, which dissolves out the chlorophyll, washed in water and placed in dilute iodine. If starch is present the leaf turns black, and if a portion of the leaf is examined microscopically the starch grains can be seen.

In carrying out the experiments several plants are put in complete darkness for twenty-four hours, after which time sample leaves are tested to see whether any starch is present. It may be said right away that starch will be absent because it has been translocated from the leaf (after conversion to glucose) during the dark period.

To demonstrate the necessity for light, one plant is exposed to normal daylight, preferably sunlight, for two or three hours, whilst a corresponding plant is put back into the dark. Alternatively a leaf can be covered with a mask of black paper in which a design can be cut as a stencil so that light can reach that part of the leaf. After exposure to light the experimental leaf (or a leaf from the exposed plant and one from the plant kept in the dark) are tested for starch by the method described, and it is found that in the light starch has appeared. In the case of the leaf with the mask the latter will appear colourless, whilst the stencilled design will be black, due to the presence of starch.

In a second experiment a leaf on the exposed plant is enclosed in a vessel containing caustic soda, so that no carbon dioxide is available to the leaf. Thus whilst most of the leaves have sunlight, carbon dioxide and chlorophyll, this particular leaf is deprived of carbon dioxide. As before, the plant is exposed to sunlight, and at the end of the experiment it is found that starch is absent from this leaf, although present in the others. Evidently therefore carbon dioxide is essential in starch synthesis.

It is not possible to extract chlorophyll without killing the leaf, and in order to demonstrate the importance of chlorophyll it is necessary to use leaves which are naturally devoid of the pigment—the leaves which are usually called variegated. Plants such as *Coleus*, *Pelargonium* and *Abutilon* are well-suited to this purpose, and if a careful drawing of the leaf is made showing the distribution of the chlorophyll, it will be found that after exposure to light etc. starch will be present only in the areas which contained chlorophyll.

Finally it can be shown that starch production is markedly reduced if the stomata are blocked, and one obvious result of such blocking will be the inability of carbon dioxide to reach the assimilating cells.

A most important feature of photosynthesis from a general biological standpoint is the fact that oxygen appears as an end product of the process. This was demonstrated long ago by Priestley, Ingen Housz and others, whilst Boussingault showed that the amount of carbon dioxide used was equal in volume to the amount of oxygen released. The oxygen which is thus released into the atmosphere largely replaces the enormous amount consumed during the respiration of living organisms. It is difficult to demonstrate simply the release of oxygen from land plants, but it can be shown fairly easily in the case of submerged aquatic plants. These of course absorb their carbon dioxide from the water, and the oxygen is released into the water, appearing as small bubbles when photosynthesis is very active. Fig. 35 shows the method by which this release of oxygen can be demonstrated. An aquatic plant such as Canadian Pondweed is put under the inverted funnel and the released gas is collected in the tube above it. It must be appreciated that the bubbles which appear are not pure oxygen, but they contain more oxygen than ordinary air.

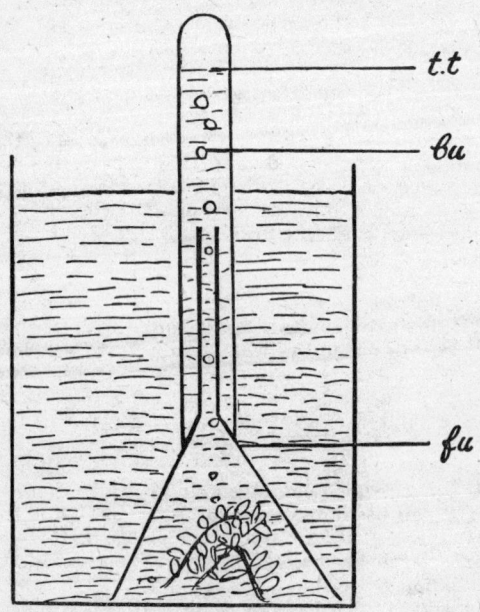

t.t

bu

fu

FIG. 35.—Experiment to Demonstrate Evolution of Oxygen from a Water Plant during Photosynthesis.

t.t. test-tube, *bu.* bubbles enriched in oxygen, *fu.* funnel.

It is usually necessary to let this experiment stand in sunlight for some time (preferably outside in direct sunlight) in order to obtain any quantity of oxygen. Under ideal conditions it is possible to collect enough gas to demonstrate the classical test whereby a glowing splint is rekindled due to the high concentration of oxygen.

These are all simple experiments which demonstrate some of the major features of photosynthesis, and modifications of them have been used to perform quantitative examination of the process. Many of these are very complex, and cannot be dealt with here.

At this stage, then, it can be said that green leaves in sunlight synthesise carbohydrate, which usually accumulates temporarily as starch. It is interesting to note that many plants which do not store starch as a reserve material do produce it in the leaves as the first bulk product of photosynthesis. How is the synthesis carried out? It is a process which has been, and still is, the subject of a great deal of investigation, representing as it does the biological source of much of the material which is used in the nutrition of living organisms—particularly man. At the present time there would seem to be considerable agreement that most of the principal stages are known, although the mechanism may not be fully worked out.

Superficially the synthesis of a sugar such as glucose may be represented by the simple equation:

$$6CO_2 + 6H_2O \longrightarrow C_6H_{12}O_6 + 6O_2$$
$$+ energy$$

but it must be emphasised that the synthesis is vastly more complex than this, and in fact there may be no direct reaction between the carbon dioxide and the water. The present-day conception of the photosynthetic process is the result of the activities of many workers over a long period of time, and briefly it may be said that the sequence of events involves an early photochemical stage which probably includes the " fixation " of the carbon dioxide and necessitates the utilisation of energy from sunlight by the chlorophyll, followed by a series of " dark " reactions which can proceed independently of the light. The idea of the " dark " reactions was put forward many years ago by Willstätter and Stoll, Blackman and others, and has been elaborated since. The difficulty has always been to establish the first organic link in the process, and a number of theories have been propounded to cover this stage. At one time it was suggested that the first substance to be synthesised was formaldehyde, which could be an intermediate product in the production of a sugar. The idea seemed feasible because of the simple balance

$$CO_2 + H_2O \longrightarrow HCHO \text{ (formaldehyde) } + O_2$$

but many factors, such as the toxicity of formaldehyde and the failure to demonstrate its presence at any stage, led to the abandonment of the idea. A more recent discovery which makes the formaldehyde idea untenable is that the oxygen produced is all derived from the water.

The modern concept of photosynthesis is a complicated one but essentially it involves the photolysis of water and the fixation of carbon dioxide. The photolytic reaction involves the separation of hydrogen and oxygen and the latter is a by-product. The hydrogen passes with carbon dioxide into a complex cycle of CO_2 fixation involving a phosphorylated derivative of the 5-carbon sugar ribulose. The activities of the cycle bring about the formation of the 3-carbon compound phosphoglyceric acid and eventually phosphoglyceraldehyde. The latter substance is the one finally converted to the typical and familiar carbohydrates glucose, sucrose and eventually starch. By other changes the phosphogylceraldehyde may be converted to fat and also to protein so that it plays a vital part in the synthesis of the main food materials. The steps in all these syntheses are very complicated and require the activities of many enzymes. It is important to remember that the main process is cyclic—simple raw materials being " fed " into a complex mechanism from which various " food " substances emerge. Only the photolysis phase requires light energy—the other cycle is kept going by energy supplied by some of its own products. Experimental work has demonstrated various phases of the process.

At this point it may be appropriate to say a little more about the substance known as chlorophyll. The pigment is known to be restricted to special cytoplasmic bodies called chloroplasts, and in some plants the green colour is masked by other pigments, such as the anthocyanin in red cabbage and the brown fucoxanthin in many seaweeds. It has long been known that in the chloroplasts there was a mixture of pigments, but Willstätter and Stoll first showed the actual composition. They showed the presence of four pigments: chlorophyll *a*, which is bluish-green; chlorophyll *b*, which is yellowish-green; the orange carotin and the yellow xanthophyll. The proportion is such that the mixture looks green, and in fact there are probably more than four pigments usually present. In the laboratory the constituents may be separated by using various solvents or by allowing the whole extract to separate on a column of absorbent powder. This latter method, known as chromatography, makes possible some very critical separations.

The extracted solutions are shown to be capable of absorbing various fractions of the visible spectrum, and absorption spectra of the chlorophyll pigments are now well known. It is this property of light absorption which makes chlorophyll the important agent in the energy transference in photosynthesis. The role of the chlorophyll seems to be as a source of electrons released by the energy of the incident light and a means of releasing hydrogen atoms from the water to be used later in the dark reaction to reduce carbon dioxide. It should also be stated at this point that the pigments vary in their contribution to the photosynthetic process. Thus in general the green pigments absorb much of the red light, with a strong band from the blue end of the spectrum. The orange and yellow pigments absorb light chiefly from the blue region. By growing plants under double-walled bell-jars which contain coloured solutions, it is found that the amount of photosynthesis is much higher in red light than in blue light. Although the utilisation of light energy is so important, not more than 1-2% of the light falling on the leaf is actually used in carbon assimilation.

The whole process is affected by various factors such as the degree of stomatal opening, freedom of diffusion in the leaf air spaces, etc. Aquatic plants often have minute leaves and chlorophyll-bearing-stems with large air spaces.

Furthermore, it is possible to demonstrate that the rate of photosynthesis is greatly affected by the concentration of any of the vital factors. Thus for a constant chlorophyll content and light intensity the rate of assimilation will increase if more carbon dioxide is made available, until a point is reached where one of the other factors is in the shortest supply and has thus become a limiting factor. In addition, temperature has a similar effect, whilst a more important consideration is the internal factor of chlorophyll and protoplasmic control. It is probably true that for a great part of the year the intensity of the light has a limiting effect.

It has been said that, in general, various products accumulate in the leaf as a result of photosynthesis, and of these the most obvious is starch. Because of the general nature of the synthesis, we can regard the leaf as being the primary and main region of the build-up of organic materials in the plant, though this role must be extended to other chlorophyll-bearing structures when they are present. This synthesis must be followed by translocation to other organs for storage or utilisation. In order that this translocation may be achieved, it is necessary that the synthesised products shall be in a soluble form, so that they can be conveyed in the sieve tubes of the phloem. The conversion into soluble substances is carried out by enzymes, of which more will be said shortly, and further changes, including the re-synthesis into storage products, is also brought about by enzyme action. It will be remembered that when the experimental plants were exposed to light, CO_2, etc., the presence of starch in the leaves was used as an indication that photosynthesis had taken place. If the plant is now put into darkness and a succession of leaves examined, it will be found that the starch gradually disappears, usually from the marginal region first. In Elodea (Canadian Pondweed) in particular the last traces of starch can be found round the base of the main vein.

Abnormal Methods of Nutrition

Although the vast majority of the higher plants obtain their organic materials in the way described above, there are some Flowering Plants which for various reasons need other sources.

In the first place, one may cite those species which possess no chlorophyll. In this respect they are of course in the same position as the Fungi and, like the Fungi, they must be parasitic or saprophytic.

There are not many Flowering Plants which are totally devoid of chlorophyll, and in Britain the principal examples are Toothwort (*Lathrea squamaria*), the Broomrapes (*Orobanche* spp.) and the Dodders (*Cuscuta* spp.) which are all parasitic on the Flowering Plants. A few Orchids as for example Bird's Nest Orchis (*Neottia*) and Coral Root (*Corallorhiza*) and finally the Yellow Bird's Nest (*Monotropa hypopithys*) are saprophytic, and are found in woods rich in humus where the Orchids at least have what are called mycorrhizal associations with fungal threads in the soil.

All the above plants are yellowish in colour, and with the exception of Dodder the only aerial portion is the flowering spike. The leaves tend to be small and simple, but Toothwort has fleshy leaves on its underground stem. Toothwort and the Broomrapes are root parasites sending out their own roots from which haustoria penetrate the vascular systems of the roots of the host plant. Toothwort commonly parasitises Elm and Hazel, but the Broomrapes have a variety of hosts. Most of them are annuals, but Toothwort is a perennial. In the case of the Broomrapes development of the seedling depends on rapid contact with the host.

Dodder is a climbing plant which must make very quick penetration into the host stem, as it has no reserves. From its stem haustoria (or " suckers ") enter the vascular bundles of the host (which may be one of a number of species, such as Heather, Nettles, Gorse, Clover, etc.) and absorb nutrient substances. When well established the plant produces masses of small pink bell-like flowers.

It should be pointed out that the term parasite involves an organism which derives its nutrient substances directly from another living organism, and though the strict interpretation of this might be very involved, there is in general a fairly well accepted idea of what the term implies in plants. Saprophytes, on the other hand, have no direct association with a living source, but obtain their materials from dead organic matter or from the non-living products of other organisms. It is probable that most Flowering Plant saprophytes can do this only through the co-operation of Fungi.

Besides the total parasites, there is a number of plants which are partially dependent on a host. The best known of these is Mistletoe (*Viscum album*), which may be found growing on a number of trees, including Apple, Lime, Hawthorn. (It is interesting to note that in this country at any rate it is rarely found on its classical host the Oak!) Mistletoe has pale green leaves, which, however, can synthesise carbohydrates. It develops into quite a woody plant with a deep wedge-shaped haustorium into the wood of the host. It is likely that Mistletoe is mainly dependent on the host for water and salts. Many more species are found in tropical and sub-tropical countries.

The other partial parasites commonly found in this country are mainly members of the family Scrophulariaceae and are parasitic on the roots of Grasses. The commonest genera are *Rhinanthus* (e.g. Yellow Rattle, which is widespread in meadows), *Pedicularis* (Lousewort), *Euphrasia* (Eyebright) and *Bartsia*. All are characterised by small yellowish-green leaves and a very meagre root system. Again they are probably dependent chiefly for water and salts, and it is quite probable that some of them can live independently.

There is another group of plants which are abnormal in their methods of obtaining nitrogen. It has already been said that the supply of nitrogen is often a difficult problem, and it appears that there are several ways of supplementing the normal sources.

The most striking examples are the insectivorous plants, in which a method of trapping insects is present, with the subsequent digestion of the softer parts of the body by enzymes secreted from the plant cells. These plants are frequently found in regions where the soil conditions operate against the normal nitrogen cycle, and there is consequently a deficiency of available nitrogen. Two genera of terrestrial plants of this type are native to Britain: *Drosera*, the Sundews, of which there are three species, and *Pinguicula*, the Butterworts, of which there are four species, though one at least is very rare. Both are found in boggy places and are small rosette plants with a poor root system.

Sundew has spatulate leaves on which stalked glands appear. Small insects alighting on the leaves (which glisten attractively because of liquid secreted on to them by the glands) are trapped by the secretions and by the movement of the glandular hairs which bend over the insect. In Butterwort the leaves are wider and sticky and there are no hairs, the insects being trapped by the stickiness. In both plants digestive enzymes are released to break down the protein, which is then absorbed through the leaf-surface. The glandular hairs of Sundew do not respond to non-protein substances.

Bladderwort (*Utricularia*) is a floating plant found in moorland pools, fenland ditches, etc. The whole plant consists of finely divided shoots which bear a number of small bladders which are probably modified leaves. At one end of each bladder is a valve, above which are sensory hairs. If the latter are touched the valve flies open, due to a release of tension, and there is a slight inward rush of water, which is sufficient to carry in microscopic animals. Within the bladder are two kinds of gland cells, one type functioning for the removal of water from the bladder so that the trap is re-set, whilst the function of the other type of cell is doubtful. It is not certain that any type of digestive enzyme is produced by the plant, and it may be that the breakdown of the animal tissues occurs through the normal processes of decay.

In tropical and subtropical regions the range of plants showing abnormal methods of nitrogen absorption is much greater— there are 200 species of Bladderwort, for instance—and some of the methods are rather more spectacular, as for instance in the Pitcher Plants. One species of Pitcher Plant (*Sarracenia purpurea*) has become naturalised and quite plentiful in the bogs of Ireland.

This topic can hardly be left without reference to the conditions found in plants of the Pea and Bean family (Leguminosae). If the roots of a Clover or Pea plant are examined it will be found that many small nodules are present on the roots. These nodules contain masses of a bacterial organism called *Bacillus radicicola*, and these bacteria are capable of fixing and utilising atmospheric nitrogen. This then becomes available by diffusion to the root of the Flowering Plant, whilst it is probable that the bacteria obtain carbohydrate, which of course they cannot synthesise. So important is this relationship that most Leguminous plants show very poor growth if the seeds are germinated in sterilised soil. This property of the Leguminous plants (or perhaps one ought to say of the bacteria) is a means of enriching the soil in nitrogen by ploughing back the crop to form a manure.

(v) Food Materials and Storage Substances

As a result of the synthetic processes various organic chemical substances appear in the plant, and whilst some of these are utilised at once, the remainder accumulate in many cases and are stored either in seeds or in vegetative organs. It is one of the features of the plant that its physiological processes are, on balance, constructive, so that the plant synthesises more material than it consumes. It is because of this that plants become valuable sources of food for animals, because the latter are unable to build up their food materials from inorganic sources.

The main groups of substances used as " food " by all living organisms are carbohydrates, fats and proteins.

Carbohydrates contain the elements carbon, hydrogen and oxygen, and include such substances as sugars, starch and cellulose. In general activity the sugar glucose, which is $C_6H_{12}O_6$, is the unit of exchange, largely because it is readily soluble. In general terms it may be said that all the carbohydrates are multiples of the glucose molecule, the molecules combining with the elimination of water (a chemical combination known as condensation). The greater the number of units involved in the combination the less soluble is the resulting substance—cane-sugar (2 units) is less soluble than glucose itself, whilst starch (50–200) is insoluble. Usually the more soluble substances are not used as storage materials—an insoluble substance is less readily attacked and can be deposited in a more concentrated form. Glucose does occur in many fruits and in such structures as the Carrot, whilst cane-sugar is found in Beetroot and the Sugar-cane. Single-unit substances are called monosaccharides, double-unit molecules like cane-sugar and maltose are disaccharides and the multiple-unit molecules are the polysaccharides. Of these, inulin is a soluble form in many bulbs (Bluebell, Daffodil, etc.), whilst starch is a common storage product in the food-grains, in Peas and Beans and in vegetative organs such as the Potato. It is hardly necessary to mention that these form major items in the diet of man.

In the Date and Lupin a form of cellulose occurs as the reserve substance, and these seeds are extremely hard.

When these reserve substances are required for growth they

are converted by enzymes into the soluble substances from which they were built up and which can be transported and utilised, and this means that in the vast majority of cases the key substance is glucose.

Starch appears characteristically in the cells as grains of various shapes. The form of the grain is often typical of the plant, and it is to a very great extent possible to identify various flours by the shape of the starch grains. The actual starch material is laid down in layers around a starting point called the hilum, and in the starch grains of Potato, for instance, the layers or lamellae are quite clearly seen. Starch gives a bluish-black colour when iodine is added, and this forms a very useful test.

Fats are substances formed by the combination of fatty acids with glycerol. They also contain only carbon, hydrogen and oxygen, but the proportion of oxygen in the molecule is much smaller, and fats represent a more economical source of energy, but one which is more difficult to utilise quickly. Fat is insoluble in water and appears as globules in the cells. It is widely distributed as a storage product in plants, occurring in seeds much more frequently than starch. Common examples of fatty seeds are found in the Sunflower, Brazil nut, Ground nut (peanut), Castor Oil seed, etc. Some of the fatty seeds are used as commercial sources of oil, particularly edible oil. Fats are also found in the structure of the plant, particularly in cuticle, where it helps to give a waterproof covering. When fats are required in metabolism they are hydrolysed by an enzyme called lipase back to glycerol and fatty acids. They can be stained black by using osmium tetroxide and give a reddish-orange reaction with Sudan III.

Proteins are substances which contain nitrogen in addition to carbon, hydrogen and oxygen, whilst in many cases sulphur is present, and in the nucleo-proteins there is also phosphorus. They are essential ingredients of protoplasm and are formed by the union of substances, called amino-acids, which may be regarded as the " unit " substances of nitrogen exchange and synthesis in the plant. About twenty-six amino-acids have been recognised as being present at some time in various proteins, and different amino-acids may be linked together to form one protein molecule. Some of these molecules consist of hundreds of amino-acid groups. Proteins are of many kinds, and probably the simplest form is albumin, which is soluble in water. The name is given to a group of substances, and one of the most familiar is egg-albumin, which is of course the " white " of the egg. Another very common form is the globulin which is soluble in dilute salt solutions and is therefore very commonly found in plant cells. Other proteins which are peculiar to plants are glutelins (which form the sticky dough when flour is mixed with water) and gliadins. Specific proteins are associated with various types of plant, e.g. leucosin is found in Wheat, zein in Maize and legumelin in Peas and Beans. In plants proteins are often found in reserve as aleurone grains, and these can readily be seen in the cells of the Castor-oil seed where they are scattered throughout the tissue, whilst in the grains of Wheat they occur in a distinct layer just under the wall of the grain and known as the aleurone layer, the rest of the cells in the storage region containing starch.

Proteins give a number of characteristic chemical reactions which can be used for identification. Some of the commonest tests are the **biuret** test, in which a protein in solution gives a violet colour in the presence of a drop of copper sulphate, followed by excess of caustic soda; the **xanthoproteid** test, where a yellow colour appears if the protein is warmed with concentrated nitric acid (e.g. the reaction seen when concentrated nitric acid is spilled on the hand) and **Millon's** test, in which a white precipitate is produced with Millon's reagent, changing to a pinkish colour on gently warming. In the xanthoproteid test and Millon's test the protein need not be in solution. Most of the reactions are due to particular linkages in the molecule.

These three classes of substances are present in all plants, and it is generally believed that carbohydrates and fats are the main source of energy for all living processes, though some are used for structural purposes. Protein is not usually used as a source of energy, but is mainly concerned in the formation of protoplasm.

Many other materials are found in plants, some of them allied to the substances already discussed, whilst others are special products. Thus gums and mucilages are found, and these are allied to carbohydrate—the gums being derived from compounds known as pentosans. From the latter also are formed the substances which give the hardness of wood and collectively known as lignin.

Allied to the fats are the waxes, the essential oils which give the odours so characteristic of some plants (Lavender, Mint, etc.), and possibly the resins, though the latter are of a very complex nature.

There are also glucosides—as for example amygdalin in Almond, Plum and others—and also the yellow, blue and red pigments found in so many plant structures. These are combinations of a sugar with some other substance and are of widespread occurrence.

Tannins are found in many parts of plants and are substances with a strong preservative action, as witness their use in leather treatment. They occur in leaves and in barks in high proportion.

Alkaloids are found in some groups of plants, and are of great importance because so many are of medicinal significance. Quinine, nicotine and cocaine are all examples of alkaloids. Many of them are very poisonous.

It is quite likely that many of the substances found deposited in the plant are of no real value to it, and may represent by-products of various reactions which are thus immobilised.

Finally we may mention substances which appear to have controlling activities rather than directly synthetic significance. Such are the vitamins of which, in most cases, the plant seems to be the ultimate source. Vitamin C from citrus fruits, the yellow pigment carotene associated with Vitamin A and Vitamin B from yeast are all common examples. There are also hormones, substances which influence growth and which have been isolated from plant organs, although there seems to be a much wider distribution in animals. These hormones are relatively unstable substances and, having been secreted, are readily destroyed, presumably by enzyme action.

(vi) Enzymes

The various substances which occur in the plant are produced from inorganic sources, simple chemical compounds like carbon dioxide, nitrates, phosphates, etc. The changes involved in these syntheses are very complex, and many of them cannot be carried out in the laboratory even with the use of the most elaborate equipment. How therefore can these various changes (and many others besides the ordinary synthetic activities) be brought about in the conditions available in the living protoplasm of the plant? Apart from other limitations the protoplasm can operate only within a rather narrow temperature range—in general, there is little activity below about 5° Centigrade, and above about 45° C. plant protoplasm is soon killed, or at any rate affected very adversely.

The answer seems to lie in the activities of substances called enzymes which are associated with all living cells. These substances are proteins, probably albumins or globulins, since they occur in solution in the cell sap, but are often associated with other materials, e.g. phosphates which act as co-enzymes.

Enzymes are *not* living substances and can be extracted from the cells in which they occur, precipitated from solution and dried, though this must be done without exposing them to much heat. If re-dissolved they are able to bring about the same changes as they did in the living tissue, though it is said that extracted enzymes are less active than those in the cells. The extracts are not necessarily very pure, but some enzymes have been prepared in a pure state.

Probably because of their protein nature they are susceptible to heat, and boiling destroys them, doubtless by coagulation of the protein. Some poisons also stop the activity of enzymes, but substances like narcotics which affect the protoplasm seem to increase the activity of enzymes, presumably because the permeability of the protoplasm is increased and there is no obstacle to the diffusion of the enzymes into the presence of the substances on which they act.

Another character is that they are affected to a considerable degree by the acidity or alkalinity of the medium (a condition which is usually denoted by the expression *pH* derived from the concentration of hydrogen ions in the solution. Neutral point is expressed by pH7, acidity being indicated down to pH1 and alkalinity up to pH14), and a particular enzyme will work only within a certain pH range.

But one of the greatest features of enzymes is that they are specific in action. A particular enzyme will not control any change, but is limited to a certain type of reaction or even to a single chemical step. Thus the enzyme invertase is responsible for the change sucrose (cane-sugar) ⇌ glucose and fructose, lipase is responsible for the general conversion of fats into fatty acids and glycerol, whilst pepsin, among others, brings about the breakdown of protein.

The method of operation of enzymes is not absolutely clear, but a fair idea can be gained in many cases. It is quite evident from experimental work that they are not affected by the changes which they bring about and can be recovered from the end-products of the reaction. It is also stated that their real activity is to speed up or slow down changes which would occur in any case. This means that enzymes do not actually initiate the reactions, but that in particular they accelerate changes which would otherwise proceed infinitely slowly. Briefly it may be said that their mode of action is to enter into temporary combination with the substance or substances concerned, bring about the changes in the molecule and separate from the end-products. This action has been likened to the fitting of a key in a lock—only one key fits the lock properly and will then turn it. The reason why one enzyme may attack several substances of a particular group is that it acts on a particular linkage in the molecule, so that allied substances possessing that link will be open to attack by that enzyme.

Another point which must be emphasised is that the direction of a reaction controlled by an enzyme is often dependent on the conditions in which it takes place. A reversal of the conditions may lead to a reversal of the direction of the reaction. So an enzyme which brings about the breakdown of a substance A into B and C may, with changed conditions, cause B and C to combine to form A.

Enzymes occur in minute amounts, and because they are not changed during their activity a small quantity of an enzyme can bring about changes in a relatively large amount of the substrate. Nevertheless during periods of very active change, as for example during the germination of seeds when large quantities of storage material are utilised, the amount of the relevant enzymes is found to increase.

Knowledge of enzyme activity and structure has been obtained from the study of both plants and animals. The enzyme called zymase, which causes the fermentation of sugar (glucose) to ethyl alcohol, and the digestive enzymes pepsin and trypsin, which bring about the breakdown of protein, have all been studied extensively for many years. In modern times the study has been much extended and has shown among other things that many of the enzymes, such as those mentioned above, are systems rather than single substances. Enzymes are classified largely by the type of action which they control. Perhaps the most familiar are the hydrolytic enzymes. These bring about changes which involve the addition or removal of water molecules during the particular reaction. In this group are the enzymes which are responsible for the breakdown of various storage substances to simpler soluble forms when the plant requires them, though again it must be remembered that the same enzymes probably catalyse the formation of the complex substances during storage. The enzyme system which is generally referred to as diastase is a typical example. This catalyses the starch $\rightleftharpoons$ maltose change and is usually accompanied by maltase, which controls the maltose $\rightleftharpoons$ glucose step. One of the simplest changes appears to be cane-sugar (sucrose) $\rightleftharpoons$ glucose and fructose in which the enzyme invertase takes part, whilst cytase or cellulase brings about the cellulose $\rightleftharpoons$ cellobiose conversion. All these are of course carbohydrate reactions

The changes involving fats are due to enzymes called

lipases, which may well be enzyme systems, bringing about the reactions fats $\rightleftharpoons$ fatty acids and glycerol.

Proteases are responsible for the various stages in the conversion of proteins into amino-acids (and vice versa) and include pepsin, which hydrolyses proteins to peptones, trypsin and erepsin (and others), which bring about the complete hydrolysis of proteins to amino-acids. There are many other hydrolytic enzymes, but those mentioned are among the ones most commonly encountered.

It is easy to illustrate some of these enzyme actions in practice. If some dry wheat is ground up with water and the resulting liquid tested for glucose, no sugar is found. If some wheat is allowed to germinate and the seedlings are then ground up with water, the liquid filtered off will contain much glucose and moreover will bring about the conversion of an ordinary starch suspension into glucose. Hence it seems fairly obvious that during germination diastase and maltase are active.

But there are other enzyme systems which are active in the plant. Thus zymase will split the glucose molecule to produce alcohol and carbon dioxide. This is one of the oldest known enzyme actions, because it is the one which takes place during the fermentation of glucose solutions by yeast in brewing, etc. In fact this zymase action consists of a long series of changes involving many enzymes, and much of the process is found in all plants in the early phases of respiration. These changes are very complex, and finally involve other enzymes which control oxidation, and are hence called oxidases.

On the other hand, during photosynthesis reactions occur which result in the release of oxygen. Here the reacting substances are reduced, and the enzymes involved are reductases, an example being catalase, which reduces hydrogen peroxide to water and oxygen.

All phases of activity in the cell including the formation of new protoplasm are controlled by enzymes. It seems fairly certain that the respiratory enzymes carry out their activities on the mitochondria, that many other enzyme systems are associated with the ribosomes, and they themselves are open to action by other enzymes. Another point which arises is that there must be some way of protecting the cell from the action of its own enzymes, such a case being found in the parasitic Fungi which are able to destroy and penetrate the cellulose walls of the higher plants. In this case it is found that the fungal walls are of a material which is apparently resistant to cytase.

It may be pertinent to point out also that since the detailed structure of many enzymes is not known, it may not be quite accurate to regard an enzyme isolated from the plant as precisely similar to one isolated from animal tissues, even though they control similar changes. Nevertheless we tend to use the same names, and it is known that animal enzymes will digest starch and other plant products.

(vii) Respiration

One of the universal characteristics of living tissues is their ability to respire—in fact when the cell ceases to respire it is dead. Respiration is a process which varies in detail in different organisms, but in the vast majority of plants it follows a fairly common pattern.

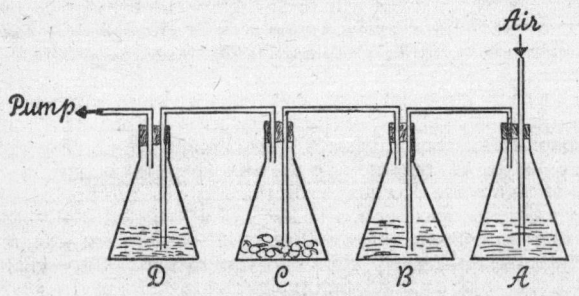

FIG. 36.—Apparatus to Demonstrate Evolution of CO_2 from Germinating Seeds.

A. Flask containing caustic soda.   B. Flask containing limewater.
C. Flask with germinating seeds.   D. Flask containing limewater.

The primary function of respiration is the release of energy to be used by the plant in all kinds of activities, such as growth, movement, chemical and physical changes, whilst some of it is lost as heat and even light. Most plants, and indeed most living organisms, take part in what is known as aerobic respiration, which means that the release of the energy ultimately involves the use of oxygen. A few plants and many bacteria and other lower organisms can obtain sufficient energy by anaerobic respiration (i.e. they respire without oxygen) to a varying degree, but it is very uncommon among the higher plants.

Some of the major features of respiration are easily demonstrated in the laboratory. If some germinating seeds are put in the apparatus shown in Fig. 36, and air is drawn through, the lime-water in the last vessel will become cloudy because carbon dioxide has been produced by the seeds and has formed calcium carbonate (which is insoluble) with the lime-water. The air is first passed through caustic soda or caustic potash to remove atmospheric carbon dioxide and then through a first vessel of lime-water to act as a control. Any plant organ could be used instead of the seeds, but the latter are particularly suited to this type of experiment because they have a high percentage of living tissue and also because at this stage they have no assimilating tissue and can therefore be exposed to light without using up the carbon dioxide which they release. If a green plant is used, the vessel in which it stands must be blacked out. From this experiment it may be seen that carbon dioxide is produced during respiration. If the apparatus in Fig. 37 is used and the carbon dioxide absorbed in caustic soda as it is evolved, it is found that the water in the long limb rises, showing that the volume of the gas remaining in the vessel is decreasing. Further observation shows that the rise of the liquid ultimately occupies one-fifth of the whole volume. Now when the cork is removed and a lighted match put in, the light goes out, indicating that the gas which has been removed was oxygen (without which combustion does not occur).

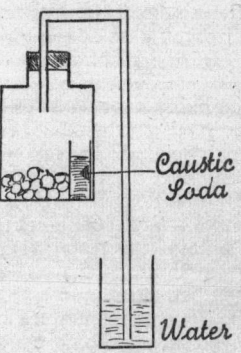

FIG. 37.—Experiment to show Absorption of Oxygen by Germinating Seeds.

So it is seen that during normal respiration carbon dioxide is evolved and oxygen is absorbed, and these are the most obvious features of the process. Other experiments can be made to show that when germinating peas are used the volume of carbon dioxide evolved is equal to the volume of oxygen absorbed. This is known as the respiratory ratio, and is characteristic of the oxidation of carbohydrates as indicated by the equation $C_6H_{12}O_6 + 6O_2 \longrightarrow 6CO_2 + 6H_2O$. All carbohydrates are respired as glucose, but this equation is complete only if we put in the energy released, which is about 725 Calories per gram-molecule of glucose and which in a straightforward oxidation would appear as heat, but in tissue respiration appears in various forms.

If, however, similar experiments are carried out with fatty seeds such as Sunflower or Castor Oil, the respiratory ratio is found to be only about 0·66—a figure which agrees with the complete oxidation of a molecule of fat (with minor variations). Actually fat is a more economical source of energy than carbohydrate and is far more widespread as a reserve, but it is not so easily oxidised. It must be remembered that here again the fat must go through a breakdown stage which probably involves glucose and that some of the oxygen is needed for that change.

Finally it may be said that in this connection protein gives a respiratory ratio of about 0·8, but it is generally agreed that proteins are not used as respiratory substrates, except when the tissue is in starvation conditions, and in any case the nitrogen part of the molecule is not used.

During respiration some of the energy released appears as heat, or even in rare cases as light. Such energy releases are probably a total loss to the plant, and in some cases the amount of heat produced is very great. Thus Fungi and Bacteria may generate heat to such an extent that in haystacks ignition may occur, and the great heat generated in manure-heaps is a familiar demonstration. In such cases the organisms are much more tolerant of higher temperatures and are called *thermophiles*. Some of the respiratory energy may be dissipated as heat or even light. Certain Fungi and Bacteria which tolerate high temperatures and are called thermophiles may cause great heating as in manure heaps. It seems likely that they may even initiate chain reactions resulting in ignition as sometimes seen in haystacks though they are killed before this stage is reached.

Organs in which there is a high proportion of living tissue show a higher rate of respiration than those in which many of the cells have died, so that the stem apex would show a higher rate of respiration per unit mass of tissue than an older part of the stem where there were dead xylem elements and fibres. If a careful analysis is made it is found that in germinating seeds and seedlings the rate of respiration is high, but as the plant gets older the overall rate falls. This is due to the increasing amount of dead tissue in the plant, because in the regions of active growth—e.g. the root and stem apices, opening flowers, etc.—the rate still remains high.

That respiration does mean the consumption of a large amount of material can be shown by a consideration of what happens during germination. If a batch of 100 pea seeds (though others will do just as well) is taken and divided into groups of ten, it is found that the *dry* weight of successive groups during germination decreases for some time, and it is only after the green leaves of the young plant start to photo-synthesise that the dry weight starts to increase again. In such an experiment it is necessary to use a sufficient number of seeds to ensure that the average dry weight of each batch would be the same. The first batch is then kept as the control, and after equal intervals of time successive groups of seedlings are removed from the soil, washed, dried and weighed. The loss of weight represents the carbon dioxide which has been evolved during respiration.

The most complex part of the respiratory process is that which occurs in the cytoplasm itself, probably mainly in the special organelles called mitochondria with their extensive in-ternal surfaces. The process has been summarised by the equation $C_6H_{12}O_6 + 6O_2 \longrightarrow 6CO_2 + 6H_2O$, but it must be emphasised that this gives little hint of the detailed changes which are involved. One of the first things that becomes apparent is that energy can be released and carbon dioxide evolved without any intake of oxygen. Much of the know-ledge of this phase came from a study of the action of yeast in sugar solutions. There glucose is broken down into ethyl alcohol and carbon dioxide even when no oxygen is available, and from this process the yeast is able to derive sufficient energy for its vital processes. It can be shown, however, that the process is very wasteful of the substrate—the reaction $C_6H_{12}O_6 \longrightarrow 2C_2H_5OH + 2CO_2$ yielding only about one-thirtieth of the energy provided by the complete oxidation of a molecule of glucose, whilst it would appear that the alcohol produced is not available for further use. Nevertheless it seems that the stages which lead to this position are of general occurrence in the plant tissues and that the initial stages of respiration are much the same whether the respiration is aerobic or anaerobic. The early phases of the breakdown of glucose are referred to as glycolysis, and involve the activity of a number of intermediate substances and enzymes. Phos-phoric acid compounds play a large part in these stages and briefly it can be said that one of the first stages is to break down the 6-carbon (hexose) molecule into two 3-carbon (triose) derivatives. Substances such as phosphoglyceric acid and phosphoglyceraldehyde appear and the reactions take place in such a way that the catalytic substances are constantly re-formed. It would appear that the pivot substance in plant respiration is a compound called pyruvic acid—a 3-carbon molecule. In anaerobic respiration this is converted to alcohol by reduction but in aerobic respiration it is oxidised to carbon dioxide and water via a complex cycle of organic acids called the citric acid or Kreb's cycle and it is in this phase that most of the energy is released. One important aspect of energy release in living tissue is that the energy is not supplied direct from these changes to the systems needing it but is " stored " in such substances as adenosine triphosphate—better known as ATP. This substance has two phosphate groups attached by what are known as " high energy bonds ". Thus in most energy requiring reactions the energy is actually supplied by the breakdown of ATP to ADP (adenosine diphosphate) and the importance of the respiratory cycle is to restore the reserve of ATP. If glucose is respired anaerobically two molecules of ATP are gained but if it is respired aerobically there is an accumulation of 38 molecules.

The ultimate role of the oxygen seems to be to oxidise hydrogen split off in the earlier stages and temporarily taken up by hydrogen acceptors such as cytochrome coenzyme. This reaction gives the water found in respiration.

The complexity of the chemical processes in respiration can be judged from the fact that fourteen intermediate stages have been recognised in the fermentation of glucose by yeast. Some indication that these stages occur in other plants is given by the fact that if germinating peas are immersed in mercury, the seed-coats having first been removed so that no air can be retained underneath, they will produce a considerable amount of carbon dioxide. This must have been produced anaerobically, but the process does not go far because after a time the peas are adversely affected by the alcohol produced.

The gases involved in respiration must move through the tissues by ordinary diffusion. The oxygen enters through the stomata and passes through the air spaces to the individual cells, the carbon dioxide leaving by the reverse route. In individual cases there may be temporary changes in the normal respiratory ratio.

(viii) Growth and Movement

Everyone is familiar with the idea that plants (and animals) grow. When seeds begin to germinate it is possible to watch the gradual development of the plant as new roots and leaves and finally flowers appear. It is also easy to see that many plants, like trees, continue to grow for many years. Here the growth is marked by phases of great activity followed by dormancy to a greater or less degree. Thus in the spring new leaves and flowers appear, whilst in the autumn the leaves fall, the fruits drop and the tree seems to become quite inactive. During the periods of activity one can see the obvious evidence of growth in the elongation of the stems and the formation of new organs. In plants the extent of growth is very varied and it is difficult to say that a particular plant will be of such and such a size; so much depends on the conditions under which the plant is growing. But the way in which growth takes place is much the same in all plants and groups of plants. There are regions in which new cells are produced, regions in which they extend either in length or width and a considerable part of the plant in which growth has ceased. The growing regions or meristems have been mentioned in connection with the anatomy of the various organs, but we are now more concerned with the general conception of growth.

It is fairly easy to observe and measure the growth in size of plant organs. Here again it is often convenient to work with seedlings because there are fewer regions to watch.

If a Broad Bean is grown in moss or loose damp sawdust it is possible to get a seedling with a straight clean radicle (the primary root). The radicle can be marked lightly with Indian-ink in equal divisions from the tip backwards. If the seedling is left to grow in a damp atmosphere and watched regularly it is found that the distance between the marks a few milli-metres behind the tip increases rapidly, but the increase in length falls off, until in the root-hair region it has practically stopped. Thus in the root the growing region is very re-stricted. If the root-tip is cut off, growth stops, which indicates that although the actual elongating region is behind the tip, it is the latter which controls the growth.

The same sort of phenomenon can be seen in the stem, but here the elongating region is much longer—perhaps a matter of feet in some vigorously growing saplings. In leaves, on the other hand, there appears to be an over-all expansion and the residual growing region is in the leaf-base—they are not apical organs.

Growth can be measured in various ways, and increase in length can very well be demonstrated by the auxanometer. This is an instrument in which one end of a fine thread is attached to the tip of the growing organ and the other end to a delicately balanced lever. By adjusting the point of balance the amount of growth can be magnified by a calculable amount at the free end of the lever and recorded on a sheet or revolving drum. By this means it can be shown that an organ does not at any time grow evenly. It may have a daily rise and fall in growth and also a seasonal rise and fall. Thus a twig does not grow steadily throughout the season even in the actual growing region. There is a period of rapid elongation in the earlier stages, followed by a gradual slowing down until elongation ceases. Presumably this is true for all the growing regions of plant, and it is paralleled to some extent by changes in respiratory rate.

Brief reference may be made to absolute rates of growth. It is usual to think of growth as a fairly slow process, although the rate does depend on many factors. Observation shows, however, that some phases of growth are very rapid, particularly in the case of plant movements. It is also true of many floral structures, and it has been shown that the stamens of some of the Grasses grow at the rate of several millimetres a day and Bamboo shoots may grow several feet in a day. This latter rate is exceptional, but it does illustrate that the process is not always as gradual as is often assumed.

Growth, however, is not merely an increase in length or even an overall increase in size. Though these phenomena are the usual obvious attributes of growth, there are other changes in the cells themselves. Thus we get differentiation from the meristematic cells leading to the various forms of mature cells. The development of the reproductive structures is also associated with growth and maturity. In general it is difficult to define growth—it is not merely an increase in size, otherwise it could be said that a balloon grows when it is blown up. It is not necessarily an increase in dry weight of the plant; no one would deny that a seed grows when it germinates to form a seedling, but it has already been explained that for a time the seedling actually falls in dry weight due to the loss of material used to provide energy. At the same time structures have appeared which were not present in the seed. Perhaps it is not necessary to try to define growth, but it has been described as an increase in size accompanied by a permanent change in form.

Growth in the plant is influenced by many factors, among which temperature, food supply, available water and various internal conditions are important. There is no doubt that a considerable part is played by the substance known as auxins or hormones, though it is not easy to see how they are transported through the tissues at the rate required to give some of the observed phenomena. The effect of auxins will be mentioned again later, but it may be said that auxin substances are used to stimulate various forms of growth, and in particular many gardeners will be familiar with the preparations used to promote root formation in cuttings, setting of fruit, etc., whilst on the other hand the comparatively recent use of various hormone preparations as weed-killers is based on the fact that the hormones produce excessive growth of the weed-plant which exhaust its reserves. The auxins are not all alike —some will inhibit growth in an organ where another will stimulate it. These growth hormones occur in very small quantities, producing results out of all proportion to the amounts used, and it is almost certain that they are rapidly destroyed, so that their effect is not continuous or cumulative. It also appears that their effects are very localised, as shown by the suppression of adjacent lateral outgrowths by a hormone released from the terminal bud. If the latter is removed the lateral buds grow out. Here the action depends on the concentration of the auxin—high concentration inhibits growth but low concentrations stimulate it. Distribution of auxins are also believed to affect leaf-fall and the fall of fruits.

In the normal way plants are not thought of as moving organisms, and the typical plant-form is one of radial symmetry which is associated with a static condition. Nevertheless plants do show many kinds of movement of their various organs. Some of these are associated with the general phases of growth, whilst others are the result of a particular stimulus.

Some simple aquatic plants are free to move by the action of fine cytoplasmic extensions called flagellae or cilia which vibrate in the water, and the male germ-cells of many kinds of plants are dependent on movement in liquid. Such movements involving a change in place of the whole unit are called *tactic* movements and are of course confined to these small structures.

Generally the plant as a whole remains fixed and its various organs move in response to different stimuli.

Most of these movements are growth movements, and consequently the results are permanent for the bit of structure involved. Any additional change must be accomplished by a further phase of growth. Thus a movement which is due to growth cannot be reversed, although its effect may be.

Some of the most familiar and widespread movements are those known as tropisms. These are movements due to an external stimulus which results in the organ affected assuming a definite position relative to the direction of the stimulus. Thus the response to gravity is called geotropism, and primary roots show a positive response, i.e. they grow towards the stimulus. On the other hand, primary stems exhibit negative geotropism. It is obvious that these responses are most important in determining the position of the organ.

Light is responsible for **phototropic** responses and produces a positive movement by stems and leaves, but roots show little reaction. The reaction of leaves is specialised; though in general they will turn towards the window in a dark room, yet they have a particular orientation, so that the blade is at right angles to the incident light. Thus they are placed in a most favourable position for their most important function—that of photosynthesis. Most flowers react positively to light, and there can be seen an even more specialised behaviour in plants like Sunflower. Here the inflorescence faces the sun constantly, and this may be called **heliotropism.**

Roots respond most vigorously to water and show positive **hydrotropism.** This will even mask their reaction to gravity if the two stimuli are not acting in the same direction. If the root-tip meets an obstacle it will curve round it.

Finally, many organs—tendrils, for instance—respond to touch, i.e. contact, and this is called **haptotropism.** The case of the tendril is most interesting. The sensitive surface lies on the inner side of the fine hooked tip with which the young tendril starts. When this comes in contact with a solid object the growth effect comes into action and the tip curves round the support. This of course increases the contact and a coil is formed.

More than one of these tropisms may be in action at one time, and it is found that some have a more profound effect than others. The case of the root and its varying response to water and gravity has already been mentioned, whilst the stem reacts more vigorously to light than to gravity. Thus if the more powerful stimulus is applied in an unusual direction it will be found that growth will occur in a direction quite different from the normal one for that organ. Thus roots will grow upward towards water.

The effect of these stimuli is most noticeable when they are applied unevenly to the organ. Thus if a Bean seedling is laid with its radicle in a horizontal position, the tip of the root will soon bend downwards, and the bending will take place in the same region as that in which growth was demonstrated. But if the radicle is rotated slowly, so that each side in turn is exposed to the stimulus of gravity, it does not bend. Again, if Wheat seedlings are grown in a pot under a cover which admits only a small ray of light at one point, then the seedlings will bend towards the light But if the seedlings are rotated so that all sides of the young shoots are illuminated in turn, then there is no bending. The seedlings of cereals are especially suitable for this type of experiment because in its younger stages the shoot is enclosed in a sheath the tip of which is particularly sensitive to light.

The movements which have just been described suggest that bending is brought about by unequal growth on opposite sides of the organ concerned, and is generally regarded as being due to the retardation of growth on the stimulated side.

If the tips of the radicle or shoots used above are removed the organ does not respond to the stimulus, but it is very interesting to note that if the tip is replaced (say, by sticking it on with gelatine), then the response is made as before, provided of course that the organ survives this treatment. In the case of the coleoptiles (the sheaths enclosing the Wheat shoots) a tiny cup of opaque material placed over the tip will also prevent light from reaching the tip, and there is no bending. Thus it seems quite clear that the tip controls the response. But it is possible to go further. Several workers have shown that if the tip is replaced unevenly, i.e. more to one side than the other, then the side on which the tip overlaps is the side on which curvature takes place, whilst an even more striking effect is obtained by standing the severed tips on agar blocks and then placing the agar block back on the shoot instead of the tip, bending taking place as before. Thus it would appear that it is something secreted by the tip which actually causes the differential growth and the secretions are the auxins already referred to.

Other theories have been advanced. It is well known that when grass stems are laid down they will bend up again from the youngest node in contact with the ground. Haberlandt investigated this, and found that in certain cells (which retain the power to grow) in the bending area there were starch grains which changed position with the angle of the stem. So he suggested that the movement of these grains when the stem fell stimulated growth of the protoplasm. This was the statolith theory.

Whatever the cause, the result is quite definite. The plant organ assumes a particular position as a result of the stimulus applied, and in general the observed effect is one favourable to the plant.

As a special case we may quote the ground-nut or peanut. This plant is not unlike a large Clover, but after flowering the fruits are pushed into the ground, and have to be thrown out in much the same way as one harvests potatoes.

The question of tropisms cannot be left without reference to a few general growth habits which are less easy to define. It can readily be shown that in the young plant the stem apex does not grow straight up, but follows a spiral path, the amplitude of this path varying, no doubt, with the rigidity of the stem. This process is called nutation, and is exaggerated in plants with weak stems, like Convolvulus and Runner Bean, and it is suggested that this arc of growth has proved of adaptive value to the plant because it is more likely to bring the stem in contact with a support. It is also suggested that this effect is a modified gravitational one and, as can be seen in the plants quoted, it continues even when the support has been encircled. It is a definite growth movement, and it may be noted that contact has no effect.

Another group of movements is that in which a response takes place without any directional reference to the appropriate stimulus. These are called nastic movements, and involve such activities as the opening and closing of flowers, the folding of certain leaves at dusk (sleep movements) and a number of other cases. In these movements a stimulus produces a movement very localised in extent, but the organ assumes a standard position, and not one relative to the direction from which the stimulus came. Several striking examples can be quoted. The Scarlet Pimpernel will close its flowers at night, or even if a big cloud covers the sun. If the sky becomes overcast Dandelions will close and also Crocuses. Although this would appear to be due to changes in the light, it seems that it is unlikely that light is the only cause. Dandelions standing in a vessel in a window in a warm room will open long before those growing outside, even though the latter may be in bright sunlight. Most nastic movements seem to be reversible, and appear to be due to changes in turgidity of certain cells. Thus at dusk or in dull weather the leaflets of Clover and Wood Sorrel fold in what is called a sleep-movement, and here the cells which undergo the rapid loss of water to bring about the collapse of the leaflet are situated in the base or pulvinus of each leaflet. Although the mechanism itself is fairly straightforward, the method of stimulation is not very clear.

A more drastic example is seen in the Sensitive Plant, where the stimulus is shock. If the leaflet of such a plant is touched, not only does it fold, but, depending on the intensity of the stimulus, the effect passes on to the whole leaf and shoot— the leaves drooping as each pulvinus collapses. After a time the organs recover. A similar shock movement may be seen in the stamens of Barberry, which fall inwards when the base is touched. This effect has an undoubted benefit in pollination, as it is likely to bring the anthers in contact with the insect which produced the stimulus.

One point which becomes clear when these various movements are studied is that there is an interval of varying length between the application of the stimulus and the response. In many of the turgor movements this interval is very short, but in the growth movements it is much longer. It is also true that once the stimulus has been registered, so to speak, the appropriate response will occur even if the stimulus is withdrawn. Thus in the experiment in which the wheat-shoots were illuminated a few minutes exposure to light would have been sufficient to have produced the curvature, even if the light had then been cut off. To some extent the degree of response is related to the intensity of the stimulus.

The effect of the various stimuli is usually to put the organ in a position which is advantageous to the plant; the orientation of leaves with the blades at right angles to the light and the ability to bend from the petiole means that they are able to make use of all the light and produces the " leaf mosaic " so well seen when looking upwards into a tree. Obviously each leaf tends to turn towards the nearest uninterrupted source of light.

4. REPRODUCTION—HOW NEW INDIVIDUALS ARE PRODUCED

(i) Flower Structure and Arrangement

Sooner or later the plant reaches a stage in its development at which it is possible for reproduction to start. This may occur within the year in an annual plant, to be followed by the death of the plant, or it may not happen until the plant is several years old, after which flowering may occur for many years in succession. The structure in which reproduction is effected in the group now studied is the flower, and it is followed by the formation of fruit and seed.

When reproduction starts certain apices in the plant begin to produce shoots which differ considerably from those which bear the ordinary leaves. These shoots will produce the flowers, and are known as **inflorescences.** Their method of growth is much the same as in an ordinary stem, and may be of a monopodial type or a sympodial form. The monopodial inflorescence or **raceme** is one in which the flowers open from below upward, so that the youngest flower is at the apex and the shoot may continue its growth. The sympodial inflorescence is called a cyme, and here the first flower to open, i.e. the oldest, is the apical one, and further growth must take place from lateral apices. In all cases the individual flowers often develop in the axils of leaf-like structures which are called bracts, whilst on the flower-stalk there may be smaller leaf-like organs called bracteoles.

The inflorescence axis is the peduncle and the flower-stalk is the pedicel. Of course there are many flowers which are solitary, such as the Snowdrop, Lesser Celandine, Daffodil and others.

Examples of racemose inflorescences are found in plants like the Lupin, Laburnum and Bluebell, and mention must be made of a special type of raceme known as the **spike**, in which the individual flowers are crowded together and are stalkless.

In plants like Buttercup, Campion and Strawberry the inflorescences are cymose. It must, however, be emphasised that many plants have inflorescences which are not readily analysed and others have very special structures. Thus the Dandelion and Daisy have complex flower-heads called **capitula**—the so-called flower of the Dandelion is actually a compact mass of small specialised flowers. Because of this

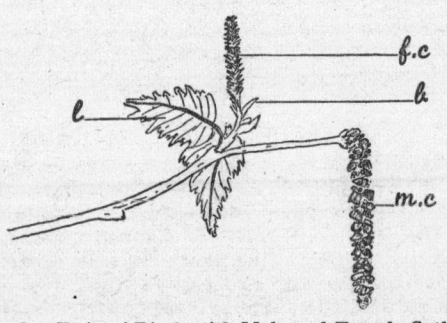

FIG. 38.—Twig of Birch with Male and Female Catkins.

b. bract, *f.c.* female catkin, *m.c.* male catkin, *l.* leaf.

the whole plant group to which these plants belong is called the Compositae. In a similar way stalked flowers may all arise from the same place on the peduncle. Such an arrangement is called an umbel, and though it does occur in many families, it is found without exception in the family Umbelliferae.

It is not possible here to discuss all the variations of inflorescence, but those which have been described will apply to a great many of the species likely to be seen.

Another modification is seen in the type of flower which is called a catkin. Those of Hazel, Willow, Poplar and others are very familiar, and consist of very simple flowers, each usually supported by a bract, arranged on an upright or drooping axis. More will be said about these flowers later. Fig. 38 shows the catkins of Birch.

The function of the inflorescence is to lift the flowers into a position in which the pollen can be transferred and received, and it will be seen that the arrangement is often associated with the method of pollination.

Although the flowering shoots frequently arise from the existing stems, there are many plants in which the whole aerial shoot is the inflorescence, the normal vegetative growth forming a rosette or group of basal leaves, whilst in many rhizomes the foliage leaves arise individually from the underground stem.

The Flower

This organ represents the ultimate phase of development in plant reproduction. In the lower plants there will be found a series of reproductive methods which show an increasing degree of complexity and efficiency. None of these can be regarded as flowers in the way in which the term is generally accepted. Many interesting ideas have been put forward about the development of the flower, but we may regard it as a shoot on which the lateral structures, and in particular the leaves, have been modified to produce the reproductive organs and also the various accessory parts which, whilst not being concerned in the actual reproduction, do help in the sequence of events which lead to this process. Not infrequently one finds freak flowers in which there are various stages of modifications between ordinary leaf-like bracts and the coloured petals, whilst in some cases the actual reproductive whorls may be replaced by leaf-like organs. Such occurrences are very suggestive of the primitive relationship.

It must be said at the outset that the structure of the flower varies a great deal in the different families of Flowering Plants. It will be quite impossible in the present book to deal with the variations, and reference can be made to only some of the general arrangements.

The upper part of the floral axis is called the receptacle, and on this the floral whorls are developed. Fig. 39 shows a simple type of flower as exemplified by the Buttercup.

Typically there are four whorls of floral parts, though one or more of these may be missing in any particular case.

If a flower such as the Buttercup is examined it will be seen that in the unopened condition it is enclosed by a whorl of simple structures which are at first green but become yellow as they open. These are the sepals, and together they constitute the calyx. In many flowers their function would appear to be protective only, and then they tend to remain

green, but in other cases the sepals become coloured and
attractive and may replace the petals. Such a condition may
be seen in an Anemone or Marsh Marigold. The individual
sepals may be free from one another or joined to form a tube,
conditions referred to as polysepalous and gamosepalous
respectively. Thus the Buttercup has polysepalous sepals,
whilst the Primrose has a gamosepalous calyx.

Within the sepals, and morphologically higher up the axis
in the simple flower, is the **corolla,** consisting of the **petals.**
Typically these are brightly coloured, and their function is

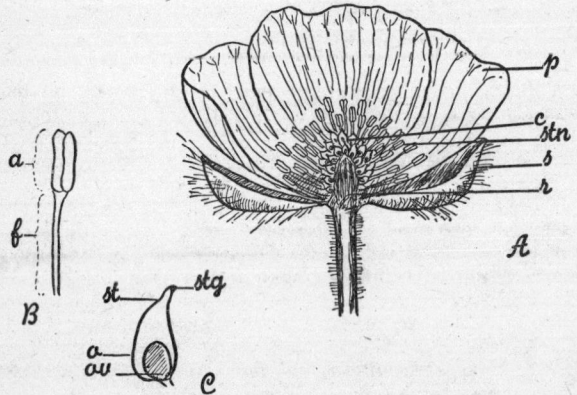

FIG. 39.—Diagram of Parts of a Buttercup Flower.

A. Longitudinal section of flower.   B. Single stamen.   C. Carpel.

a. anther, *c.* carpel, *f.* filament, *o.* ovary, *ov.* ovule, *p.* petal, *r.*
receptacle, *s.* sepal, *st.* style, *stg.* stigma, *stn.* stamen.

undoubtedly associated with the attraction of insects for
pollination. As with the sepals, the petals may be separate
or they may form a tube—polypetalous or gamopetalous. It
is the corolla which gives the characteristic appearance of the
flower, and usually we recognise a flower because of the shape,
arrangement and colour of the petals, particular emphasis
being paid to the symmetry of the corolla.

The calyx and corolla together form the **perianth,** and in
flowers such as the Bluebell, Tulip and others the two whorls
are not distinguishable except by their position on the axis.
In all flowers these floral parts are known as the accessory
whorls and are not directly concerned in reproduction, so that

after pollination and fertilisation it is found in a great many cases that the petals in particular wither and fall off, though the calyx quite often persists and is present with the fruit.

Further towards the apex of the axis—or the middle of the flower—we come to a whorl of organs which are no longer leaf-like. These are the stamens, forming the whorl known as the **androecium** and representing the male part of the flower. Typically each stamen has a stalk or **filament,** at the free end of which is a bilobed **anther,** each lobe containing two **pollen sacs** in which the pollen grains are developed. It is from these pollen grains that the male cells are ultimately produced. The stamens vary greatly in number, and whilst some simple flowers have only one each, others, like our example the Butter-cup, have scores of stamens. Not infrequently, especially when the petals form a tube, the stamens are attached to the petals (Primrose, Deadnettle, etc.), whilst in some flowers the stamens themselves may form a tube either by union of the filaments (adelphous) or of the anthers (syngenesious). The adelphous condition is readily seen in the Sweet Pea and the syngenesious state in the Dandelion family. When the pollen sacs are mature they usually open by longitudinal slits to release the pollen, but in a few cases the latter escapes through apical pores.

Finally the innermost whorl consists of the **carpels** and is called the **gynaecium.** It produces the female organs of the flower. Typically each carpel has a basal region or **ovary** in which the future seeds or **ovules** are located. From the ovary there grows up a region called the **style,** which varies considerably in length and ends in the receptive part of the organ, which is called the **stigma.** The gynaecium shows much variety in development in different families and its modifications are used a good deal in classification. It is usually regarded as the apical whorl, but modification of this may occur, and the carpels are in many cases actually enclosed in the receptacle, with the styles and stigmas protruding (Apple, Gooseberry). Also there is much variety in number and association of the carpels; thus in a Pea flower there is a single carpel with a row of ovules along one edge, in the Butter-cup there are many carpels which are all free from one another and each bearing a single ovule, whilst in the Primrose there is a single-chambered structure with a large number of ovules which can be recognised as having been derived from five carpels only by careful study of its development and by ex-amination of the fruit. The carpellary organisation may vary within a single family, and carpels may be found showing various stages of fusion. A gynaecium in which the carpels are free from one another is called apocarpous, whilst one in which they are fused to be syncarpous. In the Bluebell and Lily three chambers are visible in the ovary, indicating that three carpels were involved, but in the Campions, Stitchworts, etc., and in the Primrose family the fusion has gone to such an extent that the individual carpels are not distinguishable. It is obvious that there is a wide range of form, and it must be borne in mind that the organisation of the gynaecium has a big influence on the structure of the fruit to which the ovary gives rise.

The arrangement of the ovules in the ovary is called the **placentation** because the region to which they are attached is the placenta, and the commoner types of placentation are marginal and parietal (on the walls), axile (on the central axis formed by the union of several carpels), free central (where the individual chamber walls have gone leaving a central pillar) and basal, which is often the condition when the carpel contains a single ovule.

This brief description of the parts of the flower will give some idea of the general characteristics, but again it must be said that there is a great variety of arrangement which will be appreciated only by the examination of many flowers.

A little must now be said about the symmetry and relationships of the floral parts.

Many flowers show general symmetry, so that if it is desired to cut the flower in half longitudinally this can be done in any plane. The Buttercup, Rose, Primrose and many others show this symmetry, and such flowers are said to be regular or **actinomorphic.** Others are symmetrical about one plane only as can be seen in the Sweet Pea, Deadnettle (Fig. 40), Antirrhinum, etc., and these are said to be **zygomorphic.** These differences in form are in many cases associated with pollination and very frequently give a striking individuality to the flower, a feature exemplified above all, perhaps, in the Orchids.

There are other general features which are useful in identification and classification. Thus it is usual to find that in the group of Flowering Plants called Monocotyledons the floral parts are present in whorls of three, whereas that arrangement is unusual in Dicotyledons, in which five is a typical basic number, although there are variations.

Brief reference must be made to the succession of floral whorls. The primitive condition is one in which the carpels are apical with the other whorls successively lower on the axis, the lowest whorl being the calyx. Such a condition is called **hypogyny**—the gynaecium is superior and the calyx inferior.

It is the condition found in the Buttercup family (Ranunculaceae), Deadnettle family (Labiatae), Foxglove family (Scrophulariaceae) and many others. In other cases the receptacle may be more or less flattened or cup-shaped, and though the gynaecium is still apical, the other whorls are arranged around rather than below it. This is **perigyny**, and though the gynaecium is described as superior, no whorl is referred to as inferior. It is a condition difficult to identify because of the range of forms, and is found in the Rose family (Rosaceae) and in the Pea family (Leguminosae). Finally a stage is found in which the gynaecium (at least the ovary) is enclosed in the receptacle and the remaining whorls are situated on top of it. This is called **epigyny**, and the gynaecium is now said to be inferior and the calyx superior. Such a condition is found in the Apple group of the Rosaceae and in all the members of the families Umbelliferae and Compositae.

Not all flowers possess all the floral whorls. Reference has already been made to the occasional suppression of the petals, but a more critical aspect is the absence of one or the other of the reproductive whorls. A flower possessing both stamens and carpels is said to be **hermaphrodite** and is the commonest condition to be encountered.

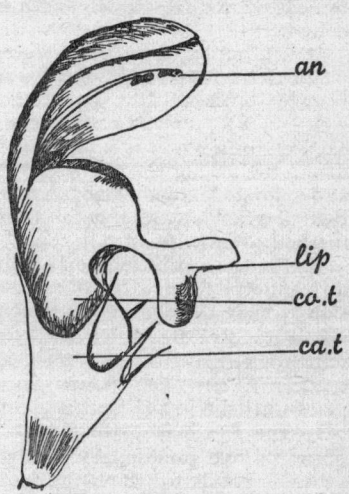

FIG. 40.—Flower of White Deadnettle.

an. anther, *ca.t.* calyx tube, *co.t.* corolla tube, *lip.* anterior lip petal.

But there are many cases in which the flower has stamens only (staminate) or carpels only (pistillate), though both accessory whorls may be present. Such flowers are said to be unisexual and may occur on the same plant, as in Hazel and Alder, in which case the plant is said to be **monoecious**, or they may be on separate plants, as in the case of the Willows, Poplars, Holly and the Campions. These species are said to be **dioecious.**

In addition to the structures already described in the flower, other organs may be present, of which the most general are nectaries. These are glands producing a sweet liquid very attractive to insects, and are usually developed in association with one or another of the floral whorls, though extra floral nectaries are not uncommon. Thus in the Buttercup family the nectaries are usually associated with the petals, and indeed the petals in some members of the family (Hellebore, Monkshood) are little else but nectaries. In other cases the nectaries are associated with the carpels. Many flowers, though perfumed, have no nectaries and are visited by insects for their pollen.

It will appear from the foregoing that the description of a particular flower in botanical terms can be a somewhat protracted process. But a number of abbreviations are in common use, together with certain conventional diagrams. To describe a flower it is usual to draw a vertical section (a line diagram showing the position of the parts on the receptacle) and a floral diagram which is a plan of the whorls and link these two diagrams by the floral formula, which is a shorthand method of description.

Fig. 41 shows this method applied to a Buttercup flower and a Deadnettle flower. In the case of the Buttercup the description is that the flower is actinomorphic ($\oplus$), hermaphrodite ($\female$), it has five free inferior sepals ($\bar{K}5$), five free petals ($C5$), numerous free stamens ($A \infty$) and numerous free superior carpels ($\underline{G} \infty$). The flower is insect-pollinated, has basal placentation and the fruit is a collection (aeterio) of achenes.

In the Deadnettle the flower is zygomorphic ($\dagger$), hermaphrodite, five gamosepalous inferior sepals ($\bar{K}(5)$), five gamopetalous petals to which the four free stamens are attached, the latter being in two whorls ($C(5)\ A2 + 2$) and two syncarpous superior carpels ($\underline{G}(2)$). The flower is insect pollinated, the placentation basal and the fruit is a carcerulus.

This method can be applied to any flower, and it is often possible to give a fairly general formula for a family or a genus in a family.

(ii) The Pollen Grain and Ovule

It has been pointed out that the essential organs in the flower, directly concerned in reproduction, are the stamens and the carpels. Within these structures develop the cells which finally unite to produce the new individual.

The stamen arises as a small protuberance from the receptacle and differentiates into the filament and anther. The latter becomes bilobed, and in each lobe two central groups of cells become distinguishable and constitute the archesporium. From this tissue the central region gives rise to the pollen mother cells, which finally divide by meiosis into tetrads—groups of four cells, each of which develops into a pollen grain. Outside the pollen grains the cell mass is called the tapetum, and this gradually breaks down and assists in the nutrition of the developing pollen grains.

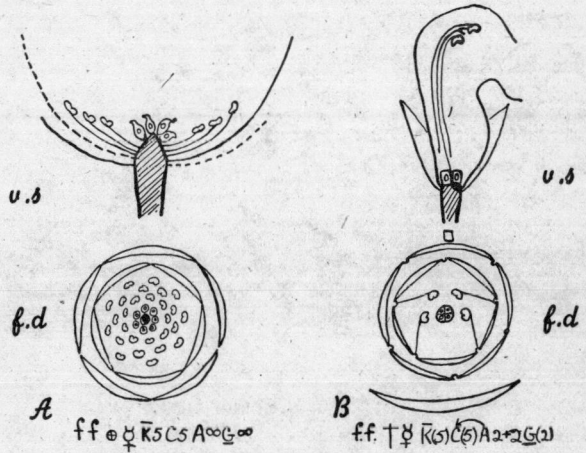

FIG. 41.—Diagrams of Floral Structure.

A. Buttercup. B. White Deadnettle.

f.f. floral formula, *f.d.* floral diagram, *v.s.* vertical section.

Immediately under the epidermis of the anther the cells outside each pollen sac become large and fibrous, except where the two pollen sacs meet. Here the epidermal cells enlarge to become thin-walled, and this is called the stomium. As the anther matures the pollen grains separate and lie loosely in the pollen sacs. Each pollen grain has at first one nucleus, which divides to produce two daughter nuclei. One of these remains surrounded by a dense mass of cytoplasm and is called the generative cell, whilst the other lies in the general cytoplasm of the grain and is called the vegetative nucleus.

Around the protoplasm two protective membranes develop—
the intine, which is very thin, and the exine, which is thick,
irregular and often waxy, so that when liberated the pollen
grains are protected against wet which would damage them.
The shape of the pollen grain is often very characteristic, so
that grains may be recognised from an individual species.

Whilst the pollen grains ripen, the anther wall changes.
The tapetum disappears and the fibrous layer becomes very
marked. The grains are liberated by the splitting of the anther

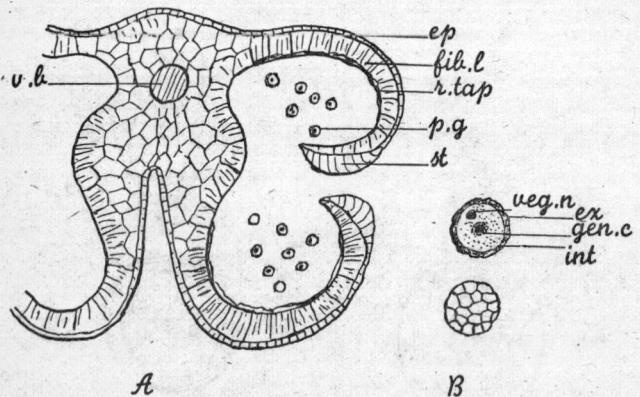

FIG. 42.—Structure of an Anther.

A. Transverse Section of Mature Anther. B. Mature Pollen Grain.

ep. epidermis, *ex.* exine, *fib.l.* fibrous layer, *gen.c.* generative cell,
int. intine, *p.g.* pollen grain, *r.tap.* remains of tapetum, *st.* stomium,
v.b. vascular bundle, *veg.n.* vegetative nucleus.

wall at the stomium—an effect which can be ascribed to the
contraction of the fibrous layer due to loss of water, with
consequent tension on the stomial cell walls, which rupture
(Fig. 42A). Pollen grains vary considerably in size, but in
general it may be said that those transferred by insects are
fairly large and often rough, whereas in wind-pollinated flowers
the pollen grains are small, very numerous and smooth.

The ovule also arises as a small cellular mass from that part
of the ovary wall known as the **placenta**. In this structure a
single cell becomes the spore mother cell, and divides by
reduction division (meiosis) into a chain of four potential

spores, but except in rare cases three of these degenerate, leaving a single cell. The surrounding tissue constitutes the **nucellus,** and at a very early stage it is possible to see the development of the two enveloping layers or **integuments** which ultimately become the seed-coats. The developing ovule may be upright, with its stalk or **raphe** at the base, horizontal, or completely bent over so that its stalk is attached to one side. At one end the developing integuments never meet, and a minute aperture is left which is called the **micropyle.** At the other end the tissue is called the chalaza.

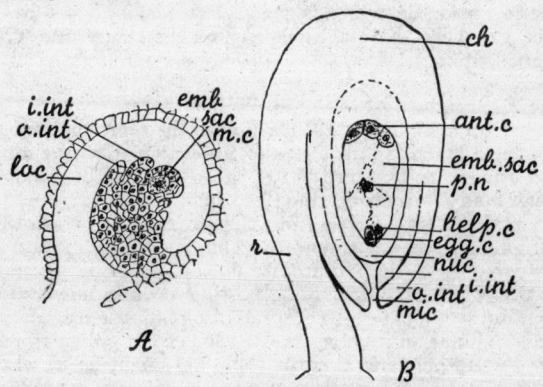

FIG. 43.—Development of the Ovule.

A. Early stage. B. Mature.

Ant.c. antipodal cells, *ch.* chalaza, *egg.c.* egg cell, *emb.sac.* embryo sac, *emb. sac. m.c.* embryo sac mother cell, *i.int.* inner integument, *help.c.* help cell, *loc.* loculus, *mic.* micropyle, *p.n.* polar nuclei, *o.int.* outer integument, *r.* raphe, *nuc.* nucellus.

The mother cell develops into an **embryo sac.** It is of some interest that the development is much later than is the case in the pollen grain. In a Bluebell, for instance, the pollen grains are fully developed early in January, whilst the flowers are still in the bulb, but the ovule may be only just completed when pollination occurs.

At first there is a single nucleus in the embryo sac (Fig. 43A), but by a series of divisions eight nuclei are produced. The original nucleus divides and the products migrate to opposite ends of the sac. Here each divides into four, and a nucleus from each end returns to the middle of the sac—the so-called polar nuclei. These may fuse to give the proendosperm nucleus. The three nuclei at the chalazal end become enclosed by cell walls and are called the **antipodal** cells. It is believed that they represent the vestiges of an alternate generation which is independently represented in the lower orders and is called the gametophyte. Of the three nuclei at the micropylar end one is actually the egg-cell and the others are called help-cells. Thus of the eight nuclei which are all potential egg-cells only one in fact functions as such. Fig. 43B is a mature ovule.

The condition described is the general one among Flowering Plants, and when this stage has been reached the ovule is ready for fertilisation.

(iii) Pollination

One of the most important features in the reproduction of the Flowering Plant is the transfer of the pollen from the anthers to the stigmas. This process is called pollination, and may involve many structural modifications in the flower. The pollen may be transferred from the stamens of one flower to the stigma of the same flower. This is self-pollination, and can occur only in hermaphrodite flowers. It does occur in a great many plants, and may be possible even in flowers which show some modification to ensure cross-pollination.

Cross-pollination involves the movement of pollen from one flower to another, and there is some argument as to whether the flowers concerned must be on two distinct plants or whether different flowers on the same plant qualify under the definition. In general pollination can be effected only between plants of the same species, but sometimes hybridisation can be effected as a result of pollination and successful fertilisation between plants of allied species. It is generally argued that cross-pollination leads to the development of seeds which produce more vigorous plants, whilst continued self-pollination may lead to a loss of vigour. Such continued self-pollination is called inbreeding. In spite of this many successful plants are almost entirely self-pollinated year after year.

Self-pollination is impossible in unisexual flowers and is prevented in many hermaphrodite flowers by a variety of devices. These include the ripening and discharge of the pollen before the stigmas of the same flower are receptive, a condition known as **protandry,** and found for example in the Dandelion and Deadnettle and the reverse condition, where the stigmas are receptive before the anthers of that flower

are ripe, as in Figwort, Plantain, Hawthorn and others, known
as **protogyny**. In other cases the stigma structure and
pollen-grain size may be incompatible, so that physical recep-
tion of the pollen is impossible. This is found in some of the
flowers of Primrose and Cowslip. In the Pansy there is a
mechanism which prevents the pollen from reaching the stigma
of the same flower. In a number of cases, as in many cultivated fruit trees, some individuals or varieties are known to be
self-sterile, and in these cases other varieties must be planted
in the vicinity so that effective fertilisation can take place.

Cross-pollination may be encouraged by various methods,
especially when insects are the pollinating agents, and a little
more will be said about this under insect pollination.

It may be said here that whilst the pollen is exposed for
pollination, it is often protected by the position of the flower
(e.g. drooping, as in Bluebell, Lily of the Valley, Cowslip), or
by closing at night and in bad weather, a familiar happening
in Lesser Celandine, Scarlet Pimpernel and Dandelion.

Insect-pollinated flowers show a much greater variety of
structure than wind-pollinated flowers, and they exhibit many
features which are designed to attract insects and which may
therefore be regarded as aids to cross-pollination. These
flowers are usually brightly coloured, often large and frequently
perfumed, although the smell may be offensive to man, especially in flowers pollinated by flies. This is true of the Wild
Arum and Moschatel. Nectar is often produced, and the floral
modifications and colour variations are frequently associated
with the location of the nectar. In many flowers the whole
structure is related to the type of insect which will be concerned
in pollination. Thus many tubular flowers, like the Honeysuckle and Tobacco, are pollinated by long-" tongued " moths,
because they only can reach the nectar at the bottom of the
corolla tube. Such flowers are also sweet-scented, and open at
night in many cases. On the other hand, open, shallow flowers
like the Buttercup, Strawberry and Stitchwort have nectar
which can be reached by short-" tongued " insects, such as
flies and beetles. Bees visit many flowers, such as Clover,
Heather, Deadnettles and other tubular flowers, from some of
which they can extract the nectar by inserting the mouth-parts,
whilst in other cases they can open, and crawl into, the flower-
tube. It is obvious that there must be a great range of adapta-
tion, and it is equally true that many insects obtain the nectar
by biting through the appropriate part of the flower.

The general principle in all cases is that the insect, in
obtaining the nectar, pushes against the stamens and so

removes pollen, which may be brushed on to the stigma of a flower subsequently visited. It is in this respect that the devices operate to prevent self-pollination. Many flowers have mechanical devices which can be operated only by large insects such as bees. Thus in the Snapdragon only a large bee can force down the lip and enter the flower. No doubt many readers will, as children, have trapped humble-bees by nipping shut the flower of the Snapdragon.

In spite of all the devices, however, it must be emphasised that cross-pollination cannot be compelled, and that many plants do depend in the last resort on self-pollination. In the case of Wood Sorrel and Violet, special flowers are produced which never open, and here the pollen tubes grow directly from the anther to the ovary. Very often these **cleistogamous** flowers, as they are called, are the only flowers to produce seeds, because the normal flowers are open at a time when few insects are about and in addition they are often concealed.

The other main pollinating agency is wind, and the flowers which are adapted to this method are easily recognised. It will readily be appreciated that the pollen should be freely released so that the wind will distribute the grains. Equally necessary is a carpellary arrangement which will trap the pollen. So it is found that wind-pollinated flowers are usually very simple structures, clustered together without any petaloid whorls, conspicuously exposed and often developed before the leaves appear or whilst they are still small. There are no nectaries or bright colours, but the stigmas are often large and feathery or sticky, whilst the anthers may be exposed at the end of long filaments. At the same time some of these flowers are easily seen by man, in spite of the lack of bright colour, simply because there is no foliage to obscure them.

Examples of wind-pollinated flowers are found in the catkins of many trees, such as Hazel, Poplar, Birch (Fig. 40), Alder, Elm, Oak, Sweet Chestnut and in other plants such as Nettles and all the Grasses. Most of the trees produce their catkins early, and many people will have seen the clouds of pollen which are blown from the Hazel catkins. The Grasses flower later in the year, but their flowering stems grow high above the general level of the foliage, and the flowers are marked by the extremely long filaments of the stamens and the much-branched stigmas. The flowers of Grasses open very rapidly.

(iv) Fertilisation

By one means or another the pollen-grain is transferred to the stigma. Here it lodges among the lobes or sticks to the surface to be joined by many others. Often the flower begins to wither after pollination, or it may close, but usually there is a fairly long period in which pollination can take place.

The stigma frequently secretes a nutritive substance which can be absorbed by the pollen-grain, and as a result of this the pollen-grain germinates. From one of the thin places in the exine the intine is pushed out by the expanding cytoplasm, and the structure so formed, the pollen-tube, grows down through the tissue of the stigma and into the ovary. The vegetative nucleus remains near the tip of the pollen-tube and the generative cell moves down, its nucleus dividing to form two male cells or gametes. Eventually the tip of the pollen-tube reaches the ovule and, as a general rule, enters the micropyle, though in some cases it penetrates the chalaza. The tip of the tube breaks down after penetrating the nucellus and the two male cells are released near the egg-cell or oosphere.

The help cells have disintegrated, and are probably associated with the penetration of the nucellus by the tube.

One male cell—the true male gamete—unites with the egg-cell to form the zygote, which will develop into the new plant. The other unites with the polar nuclei (proendosperm nucleus) and stimulates rapid nuclear division, the latter often outstripping cell-wall formation. This cellular mass is the endosperm, and is the ultimate source of nutrition for the developing embryo.

This " double fertilisation " is characteristic of Flowering Plants, and it must be emphasised that in this group the endosperm is a post-fertilisation development.

(v) Development of the Seed and Fruit

As a result of fertilisation many changes take place in the fertilised ovule and the ovary. In general, the petals, and often the sepals, are already withering, and the stamens have shrivelled soon after their pollen has been shed.

The carpel or carpels, and in particular the ovary, now become the most conspicuous part of the flower.

The actual sequence of events may vary somewhat in different cases. This is very obvious in the wide range of fruit types which can be seen, but it is also true in seed formation, and though the general pattern of development is similar in the Dicotyledons, there is greater variation in Monocotyledons.

The true fruit is produced from the ovary, whilst the seed is the fertilised ovule. It is not absolutely clear whether the formation of a fruit is dependent on fertilisation. Many cultivated fruits are seedless and a few wild ones are sterile.

It may be that pollination provides the necessary stimulus, or there may be fertilisation without seed production. Except in rare cases, however, fertilisation is necessary to produce a fertile seed. Cases of seed formation without fertilisation are called parthenogenesis, and do occur infrequently in Flowering Plants. The oosphere or egg-cell which develops in this way has not been formed by reduction division. Parthenogenetic development is found in the Dandelion.

The Seed

As was previously suggested, the further development of the seed may vary somewhat in different cases, but the procedure to be described is a fairly general one. Stages in development can be isolated quite readily from the developing fruits of Shepherd's Purse. If these are gently squeezed on a microscope slide the young embryos can be pressed out.

Immediately after fertilisation the zygote or oospore divides transversely into a basal cell (which undergoes no further development) and an upper cell. The latter divides again into a suspensor cell (which is thus the middle cell) and a terminal embryonal cell. The suspensor produces a chain of cells, and this elongation pushes the embryonal cell upwards towards the endosperm mass which has been forming. The embryonal cell divides initially into eight cells (the octants). The four anterior ones produce the embryo shoot or plumule and the cotyledons, the four posterior octants produce the hypocotyl, and the terminal cell of the suspensor gives rise to the young root—the radicle. Gradually the embryo takes the shape of a tiny plant, and further development may take place in different ways. In plants like Peas and Beans, the Sunflower and many others, the material which has accumulated in the endosperm is transferred to the cotyledons or " seed leaves ", which thus become swollen and are the storage organs in the mature seed. In many cereals and in the Ash and Castor Oil the endosperm remains as a mass of reserve material, and in the Castor Oil it encloses the cotyledons which are thin and leaf-like. The former type is said to be non-endospermic or exalbuminous, whilst the latter are endospermic or albuminous. During this development other changes have been taking place. The whole seed has of course enlarged and the seed-coat have differentiated into protective layers. These may be fused, as in Pea and Bean, or separately recognisable, as in Vegetable Marrow. The outer coat is the testa and the inner one the tegmen.

Their differentiation reduces the micropyle to a narrow tube, but it is not obliterated. The whole seed gradually dries as it matures, and the loss of water helps a great deal in the preservation of the seed. The testa is usually tough and often waxy, especially if the individual seed is released from the fruit.

Thus when mature the seed consists of an embryo, a reserve of food material and protective coverings. The degree of differentiation of the embryo varies a great deal. Thus in the Pea and many cereals the seed will germinate as soon as it is mature (in wet weather Peas will often germinate in the pod and Wheat-grains in the ear), whilst in other cases, e.g. Lesser Celandine, the embryo is relatively undeveloped when the seed is shed and a maturation period is necessary—usually the process is complete by the following spring. In Orchids, however, it may be two to three years before the seed is ready to germinate.

The Fruit

During the period of seed development the ovary wall undergoes changes which lead to the appearance of the fruit. These changes vary enormously, as can be found by examining the different fruits, but in all cases there is a considerable enlargement, often quite out of proportion to the size of the seed. Not unusually other parts of the floral axis are included in the fruit, and the term " false fruit " is sometimes used to describe these structures. A case in point is the Strawberry, where the fleshy part is the receptacle and the true fruits are the tiny one-seeded " pips " on the surface.

The function of the fruit is the protection of the seed, especially during development, but very often it aids considerably in the dispersal of the seeds. The details of fruit development are related to the way in which these functions are carried out, and reference must now be made to the various types of fruit which may be encountered.

The popular conception of a fruit is that of a succulent structure which can be eaten (by man or other animals), but such forms are only a fraction of the total. Fruits are classified in many ways, but there is a general distinction between fruits with a dry wall and those with a fleshy one.

The wall of the fruit is derived from the ovary wall and is called the pericarp. It may be recognisable only as a single composite layer, or it may be separated into several regions known as epicarp, mesocarp and endocarp. In dry fruits there is a distinction between one-seeded or **achenial** fruits (which do not open) and many-seeded fruits which usually split to release the seeds and which are called **capsular** fruits. In some cases the many-seeded fruits split into one-seeded portions—**schizocarpic** fruits. The ultimate result is the dispersal or scattering of the seeds so that each one has a chance of developing free from the competition of those which were produced with it.

The structure of the fruit, and indeed the seed, is often closely bound up with the method of dispersal, and classification of fruits is often based on this consideration.

Thus succulent fruits are dispersed mainly by animals. The latter eat the fleshy part of the fruit, and the seed, which may be protected either by a hard region of the pericarp or by a hard seed-coat (testa), is rejected by the animal or passes undamaged through its alimentary canal. In the Plum and Cherry the " stone " is the hard endocarp of the fruit wall, whilst in Apple and Tomato each seed has a hard testa. It is true that some fruits are taken because of the food value of the seed (Hazel, Oak, Beech, Walnut), and here the successful distribution and germination of the seed depend on some seeds being lost when the animal is carrying them away.

The common types of succulent fruit are:

(1) The drupe, which is the fruit of the Plum, Cherry, etc., whilst the Blackberry and Raspberry have collections of drupels. The pericarp has a membranous skin (epicarp), a fleshy mesocarp and the stony endocarp protecting the seed. Exceptional types are the Coconut, with a fibrous mesocarp which helps in distribution by acting as a float, and the Walnut and Almond, with fibrous or membranous mesocarp. Drupes are usually one-seeded.

(2) The berry—a fruit in which the seeds (usually several except in the Date) lie freely in the fleshy mesocarp. Examples are seen in the Gooseberry, Grape, Banana, Orange and Cucumber. The term berry is often used for fruits which botanically are not berries. Fig. 44 is the Tomato.

(3) The pome. This is really a false fruit, because the fleshy part is the receptacle which encloses the membranous wall of the true fruit. Examples are seen in the Apple, Fig. 45, and Pear, where the true fruit is the five-chambered " core " with its rows of seeds. Hawthorn is a pome with a single seed. In these cases the fruit is developed from an inferior gynaecium on which the other floral parts were arranged.

Other individual cases of succulent fruits are seen in the Strawberry and Rose, but they are not general types. In Fig and Mulberry there are complex fruits which develop from several flowers. In addition, there are some fleshy capsules.

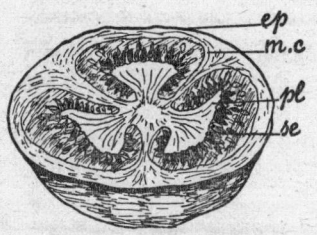

FIG. 44.—Cross-Section of Berry (Tomato).

ep. epicarp, *m.c.* mesocarp, *pl.* placenta, *se.* seeds.

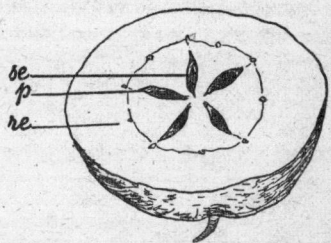

FIG. 45.—Cross-section of Pome (Apple).

p. pericarp, *re.* receptacle, *se.* seed.

Dry fruits can be classed in three groups. In the first group are those which have developed from a single carpel containing a single ovule and which do not open to shed the seed. These fruits are called **achenial**, and are of various forms, differentiated partly on structure and partly on dispersal. The term

FIG. 46.—Example of a Nut—the Acorn of Oak.

A. External Appearance. B. Vertical Section.

cu. cupule, *cot.* cotyledon, *n.* nut, *p.* pericarp, *ra.* radicle.

achene is a general one, but certain well-defined types are mentioned below.

(1) The *nut.* Seed enclosed within a woody pericarp which usually has a cupule, e.g. Hazel, Oak (Fig. 46) and Beech.

(2) The **cypsela.** Fruit usually surmounted by a parachute (pappus) of hairs. Typical of Dandelion family.

(3) The **samara** (Fig. 47). Pericarp extended into a thin wing, e.g. Sycamore (Fig. 49), Ash, Elm.

(4) The **caryopsis** which is the characteristic fruit of Grasses and is well seen in Wheat, Oats, Maize, etc. The fruit wall and seed wall are fused and the fruit is enclosed in bracts.

Fruits which contain many seeds may have developed from one or more carpels and may retain the divisions of the syncarpous gynaecium, or these may have broken down.

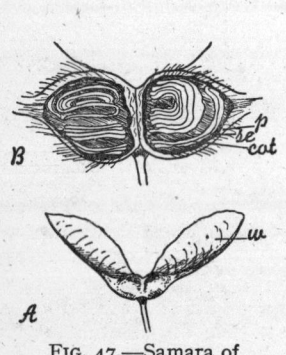

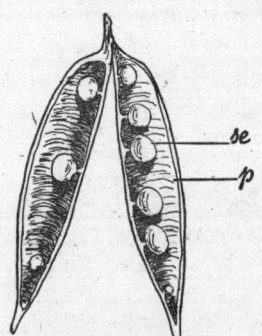

FIG. 47.—Samara of Sycamore.

A. External Appearance.
B. Vertical Section.

cot. cotyledon, *p.* pericarp,
se. seed, *w.* wing.

FIG. 48.—Legume of Garden Pea.

p. pericarp, *se.* seed.

These fruits usually open or **dehisce** to release the seeds, and are often classified according to the method of opening.

The typical fruit of the Pea family is the **legume** derived from a single carpel and opening along both edges to release several seeds (Fig. 48). In Aquilegia, Marsh Marigold and others the carpel develops into a fruit opening along one edge only, and this is called **a follicle.** In Wallflower, Cabbage and other plants of the family Cruciferae the pod is usually long and opens by two valves from below upwards, exposing the seeds on a median partition. This is the **siliqua.** Finally there is a whole group of multicarpellary fruits which open by a variety of methods such as apical teeth (Campions, Primrose), longitudinal valves (Iris, Foxglove), pores (Snapdragon and Poppy) (Fig. 49) or a cap as in the Scarlet Pimpernel and Henbane. All these fruits are referred to as **capsules**.

The last group of fruits are those which split into one-seeded portions and are called **schizocarps**. Examples are seen in the Radish, where a pod-like fruit is divided by transverse constrictions and is called a **lomentun**, in the Geranium family, where the fruit is a **regma** in which each one-seeded portion or coccus opens to shed the seed, and in the whole family of Umbelliferae (Parsley, Carrot, Hogsweed (Fig. 50)), where each fruit splits into two one-seeded halves, the whole fruit being called a **cremocarp**. In the Deadnettle family the fruit has

FIG. 49.—Capsule of Poppy.

p. pericarp, *po*. pore.

FIG. 50.—A Schizocarpic Fruit—the Cremocarp of Hogsweed.

mcp. mericarp (half fruit), *v.vitta* (oil duct).

two carpels, but splits into four one-seeded nutlets. This is a **carcerulus**.

Hence we see that there is a great variety of fruit forms, but some are sufficiently characteristic as to be regarded as diagnostic of a particular family.

From the various types of fruit we should endeavour to trace the adaptations for disperal. Successful dispersal over a wide area increases the chances of a seed reaching suitable ground for growth and decreases competition from the other seeds and the parent plant. Naturally many of the seeds are wasted, but the Flowering Plant usually produces far more seeds than are necessary to maintain its numbers.

The main methods of dispersal of fruits and seeds are by animals, by wind and by mechanical distribution, such as the explosion of the fruit. Capsular fruits open to release the seeds, and these may or may not have accessory methods of dispersal.

Dispersal by animals is either voluntary or involuntary. In the former case the animal takes the fruit for food and the seed is either rejected or passes through the alimentary canal undamaged. In such instances the seeds are scattered but when the seeds are taken for food, only those which are lost, or buried by the animal and forgotten, have a chance of germinating.

Involuntary or accidental dispersal is due to the attachment of the fruit or seed to the coat of the animal. This is made possible by the development of hooks from the surface of the fruit or seed, and these become attached to the animal when it brushes past the plant. Later the fruit may be rubbed off as the animal moves about. Most of the examples are achenial fruits and can be seen in Goose-grass, Agrimony and Bur Marigold. In the Common Burdock the hooks are actually bracts enclosing the one-seeded fruits.

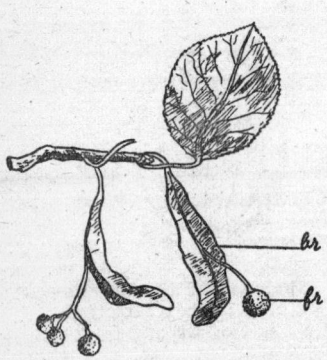

FIG. 51.—Winged Achene of Lime.
br. bract, *fr*. fruit.

Wind dispersal is achieved by such developments as the hairy parachute in Dandelion, Groundsel and Thistle, the feathery styles of Clematis and a special hairy outgrowth on the **seeds** of Willow and Poplar. In Ash, Elm and Sycamore the fruit-wall expands into a membranous wing, and in Lime (Fig. 51) and Hornbeam the same result is achieved by a large bract on the fruit-stalk. A few seeds such as those of Orchids are light enough to be blown about without any aids, whilst it must be remembered that in many capsular fruits (Poppy, Campion, Bluebell, etc.) the swaying of the stem in the wind flings out the ripe seeds. This is often referred to as a censer mechanism.

Mechanical distribution is of various kinds, but usually depends on drying of the fruit-wall in such a way that it produces a sudden split along a line of unthickened cells. With the release of tension the valves of the fruit frequently twist violently, flinging out the seeds. This can be seen in Gorse and Lupin, and frequently in some of the Cruciferous plants, such as Wallflower. On the other hand, the explosion may be due to excessive turgor in some of the cells, so that a touch causes a break in the structure and again the seeds are flung out. One of the most spectacular examples is the Balsam, commonly found along river-banks.

A few fruits and seeds are dispersed by water, the classical example being the Coconut, supported by its fibrous mesocarp.

After the seed has been dispersed it normally passes through a period of dormancy, and in some cases this is necessary for final development of the embryo.

It is because of this ability to remain dormant during adverse growth conditions that the seed has given to the Spermatophyta the dominant position they hold among plants. Seeds may remain dormant for many years without losing their ability to germinate. This is particularly true of many cereal grains, but of course the question of germination depends on factors which will shortly be discussed. Some seeds, as for example Peas and Beans, frequently lose all power of germination after two or three years.

(vi) Germination

It has already been said that the mature seed consists of an embryo, a reserve of food either in the endosperm or in the cotyledons and a protective coat. In this form it normally remains in the soil until conditions are favourable for the beginning of growth, but in seeds which rapidly attain maturity the onset of favourable conditions may induce germination even before the seeds are shed. This is unfortunately all too easily seen in Peas and Wheat, etc., in a wet season. Examination of such seeds shows that the embryo is well developed when the fruit is ripe.

Under normal conditions the coincident operation of a number of factors in the soil is necessary to stimulate germination. In order to grow, the embryo needs water, food material in a form which can be absorbed, a supply of oxygen and a suitable temperature. In this country the soil usually contains plenty of water and an adequate supply of oxygen (unless the soil is waterlogged), but during the winter the temperature is too low to permit of water uptake, and so development is held up. In desert soils there may be plenty of oxygen and a suitable temperature, but insufficient water until the rainy season. In waterlogged soils or in the mud of river-beds seeds have remained for years and failed to germinate because of the lack of oxygen. When the river was dredged and the mud thrown out on the bank the seeds soon germinated.

As the temperature of the soil rises in the spring, the seeds begin to absorb water, much of it through the micropyle, but some of it through the seed-coat. As the water is absorbed the reserve materials are hydrolysed under the influence of enzymes, and this action gradually speeds up. As the soluble products pass into solution the osmotic pressure rises and more water enters. The cells of the embryonic tissue expand and the young root and shoot begin to enlarge. Gradually the

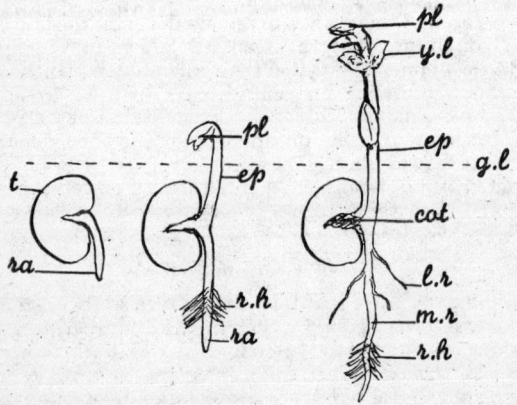

FIG. 52.—Stages in Germination of Broad Bean (Hypogeal).

cot. cotyledon, *ep.* epicotyl, *g.l.* ground level, *l.r.* lateral root, *m.r.* main root, *pl.* plumule, *ra.* radicle, *r.h.* root hair, *t.* testa, *y.l.* young leaf.

seed-coat is split, and the first part of the new plant to emerge is usually the primary root or radicle, which pushes down into the soil. More and more reserve material is brought into solution, new cells are formed at the apices and the rate of growth quickens.

The method of germination differs somewhat in different seeds.

In the Pea, Broad Bean (Fig. 52) and many others the food material is stored in cotyledons, and as it is brought into solution the soluble materials are transferred to the young root and shoot. The latter, known as the plumule, also expands and pushes upwards through the soil. The lengthening portion of the stem is called the epicotyl, and this process leaves the cotyledons in the ground. This type of germination is called hypogeal, and the first leaves to appear are true leaves. It may be observed that the young shoot does not grow upright through the ground, but is hooked so that the delicate leaves are protected. On reaching the surface the plumule straightens out and the leaves unfold. As growth continues secondary roots appear and an extensive root-system gradually develops. If the primary root or radicle persists and enlarges it frequently produces a tap root, but if it dies back, then it is usual to find a fibrous root system.

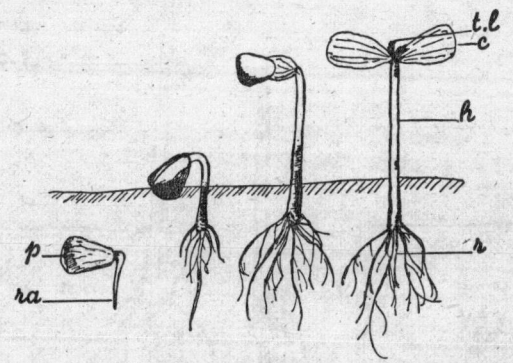

Fig. 53.—Stages in Germination of Sunflower (Epigeal).

c. cotyledon (now functioning as foliage leaf), *h.* hypocotyl, *p.* pericarp, *r.* roots, *ra.* radicle, *t.l.* true leaf.

In Sunflower (Fig. 53), Kidney Bean, Sycamore and many others the procedure is different. After the radicle has appeared, that part of the axis below the cotyledons elongates, and so the cotyledons are pushed above the ground, where they frequently turn green and act as the first foliage leaves. This type of germination is called **epigeal**, and in such plants the plumule is usually much less well developed than in hypogeal types. In both the types of germination quoted the seeds were without endosperm, but in Castor Oil and Ash the cotyledons come above the ground enclosed in the endosperm, and presumably some absorption must take place by diffusion through the cotyledon surface. Here again the cotyledons become the first foliage leaves.

In the cereals such as Wheat or Maize the procedure is different again. Here the embryo has a well-defined radicle and plumule and is buried in the single cotyledon and the food material is in the endosperm. When germination commences the radicle emerges in a sheath or **coleorhiza,** which bursts almost immediately and the radicle grows out. This is seen in Fig. 54. The plumule also emerges from the cotyledon enclosed in a sheath, the **coleoptile,** and this persists unbroken until the tip is some distance above the ground, but eventually

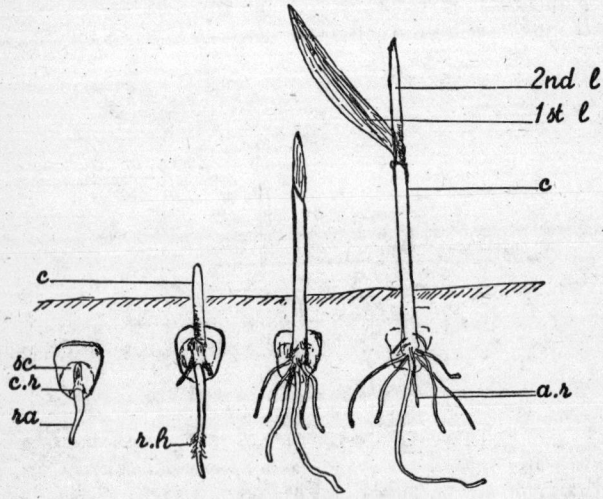

Fig. 54.—Stages in Germination of Maize.

a.r. adventitious roots, *c.* coleoptile, *ra.* radicle, *r.h.* root hair, *sc.* scutellum, *1st.l.* first leaf, *2nd.l.* second leaf, *c.r.* coleorhiza.

the first leaf pushes through. The actual stem is quite short and the leaves are rolled. As the first one opens out, the second one is revealed, and so the leaves appear in succession.

From the base of the shoot adventitious roots appear, and branching of the original root is much less marked than in Dicotyledons. In Monocotyledon seeds of this type the cotyledon appears to act as an absorbing and transferring agent between the endosperm and the embryo. Other types of germination may be encountered, but the above are representative.

In all cases the progress of germination is marked by rapid use of the reserve materials both for the formation of new tissue and for the production of energy for growth. As a result the dry weight of the seedling falls in spite of its obvious development. Normally the fall is halted when the new leaves begin photosynthesis, but if the seedlings are grown in the dark they become elongated, remain yellowish (etiolated) and the leaves are small. Because they do not lose water, such seedlings become very attenuated, but eventually they collapse as the food reserves are exhausted. Under ordinary conditions the reserve of material is more than sufficient to maintain the plant until photosynthesis starts. This does not actually occur as soon as the leaves unfold—there is a development period even after chlorophyll is present. The young leaves of many Dicotyledon plants are often of simpler structure than the adult ones and show what is called a juvenile form. Successive leaves form a series of stages towards the adult structure.

When the leaves are actually synthesising, the plant can be said to be fully established, and the further development of the organs will take place along the lines indicated in earlier chapters. Of course many seeds may germinate in situations which will not permit the young plant to survive to maturity, and no doubt others are smothered by the competition of surrounding vegetation, whilst many young plants will be destroyed by grazing, etc. Usually, however, a sufficient number of individuals survive to ensure the continuation of the species.

3

NON-FLOWERING PLANTS

So far attention has been confined to the structure and behaviour of the flowering plant. Now we must consider some of the other kinds of plants which help to make up the vegetable kingdom. Some of these plants are familiar and quite conspicuous, others are not easily observed and often difficult to identify when found. In general, they show a gradual increase in complexity of organisation both vegetatively and in reproductive development, but it is not easy to find direct links between one group and another. The result is that we can pick out several groups of plants which have fairly well-defined features distinguishing them from other groups. At the same time there may be many points of similarity, and it is on these points, together with the fossil links, that we base the idea of the gradual evolution of the higher plants.

Briefly the non-flowering plants may be classified in the following sequence.

I. Thallophyta:
    (1) Algae   Chlorophyceae (green)
                           Phaeophyceae (brown)
                           Rhodophyceae (red)
                           Cyanophyceae (blue-green)
    (2) Fungi
    (3) Bacteria (included at this stage but actually their position is not easy to define).

II. Bryophyta:
    (1) Hepaticae   Liverworts
    (2) Musci   Mosses

III. Pteridophyta:
    (1) Filicales   Ferns
    (2) Equisetales   Horsetails
    (3) Selaginales   Selaginella
    (4) Lycopodiales   Clubmosses

IV. Gymnosperms:
    (1) Coniferales   Pines, Yews, Spruces, Cedars, etc.
    (2) Cycadales   Cycas
    (3) Gnetales   Gnetum, Ephedra

It will not be possible to deal in detail with the structure and development of all these groups, but the essential features of representatives of the major divisions will be discussed so that their organisation can be compared with that of the Flowering Plants. It is probably true to say that there is a much greater variety of form in the lower groups than in the more advanced ones, and in the lower groups it is therefore more difficult to select an example which will be representative of the whole section.

THALLOPHYTA

(i) The Algae

The Algae show a wide variety of organisation within certain structural limits. At the lower end of vegetative structure the plant body is a single cell, whilst in their most developed form the Algae include species over 100 feet in length. A great majority of the species are aquatic, both fresh-water and marine, and none of them can withstand desiccation, so that even though many types are found on the surface of the ground, they are limited to very moist environments.

All the Algae possess chlorophyll, but in many of them it is masked by other pigments, though these do not prevent the plants from carrying out photosynthesis. The largest and most conspicuous Algae are the Phaeophyceae or Brown Algae and the Rhodophyceae or Red Algae. Both are almost entirely marine with the Red Algae occurring in general in deeper water than the Brown. In these groups the plants often show a considerable differentiation of the body into what is called a thallus.

The thallus is a cellular organisation which may show much variety of general form, but not a great deal of internal differentiation. There are no roots, stems or leaves, but the thallus frequently has an attaching region called the holdfast, a stalk-like part called the stipe and then the flattened " frond ".

None of the Algae has any true differentiated conducting tissue in the sense of vascular elements, and there are no lignified cells. It has been suggested that some of the cells in the larger types become modified to act as transporting elements, but their actual status is doubtful. Throughout the Algae the cell-wall is of a cellulose nature, but with a good deal of mucilage which possibly helps to give pliability to the aquatic forms, and certainly helps to prevent desiccation when the thallus is exposed to the air.

Both **sexual** and **asexual** methods of reproduction are found, and it may be convenient at this point to explain what is meant by those terms.

Sexual reproduction, as we have seen in the Flowering Plant, involves the union of two gametes or germ-cells. These are usually distinctively male or female, but in some of the lower plants there is no obvious difference. Normally, however, the male cell is a spermatozoid and is motile, so that water becomes necessary for fertilisation to be effected. This is not the case in the seed-plants and has had much to do with their present dominance. The female cell is an oosphere and is usually larger and inactive; indeed in most cases it is retained on the plant. Typically the gametes are single cells, and are produced in structures called gametangia, although the latter term is often superseded when the gametangium has a specialised form. The cell produced by the union of the gametes is a zygote.

Asexual reproduction is usually achieved by a single individual without any union of cells. It may involve the production of a special reproductive structure or **spore**, but the term may also be extended to include the separation of vegetative parts which can carry on independent growth. This, however, is often referred to as vegetative propagation, but whereas it is easy to differentiate the methods in a highly developed plant, it is not so easy to separate them in a unicellular individual.

Both forms of reproduction are found in the Algae, and in many cases the asexual reproductive structure is a motile body called a **zoospore** which swims actively for at least a short time. The sexually produced zygote frequently develops a protective wall, and is then a **zygospore**, and may remain dormant for some time (usually until growth conditions are again favourable). In other cases it may start to develop straight away. In the Red Algae there is frequently a complicated sequence of events in the reproductive cycle.

By examining a few examples of the group we shall be able to get a general idea of the characteristic feature of the Algae.

The unicellular forms are quite common in fresh water and sometimes occur in great numbers even in casual pools, a state of affairs which is most likely due to the distribution of the resistant zygospores. A common example is Chlamydomonas, of which there are several species, whilst many other forms occur which are quite similar. The plant consists of an ovoid cell with a mucilaginous cellulose wall enclosing the cytoplasm. From the more pointed end a pair of fine cytoplasmic strands

protrude, and these are known as **cilia** or **flagellae.** They
are vibratile and assist in the locomotion of the cell. Within
the wall the protoplast encloses a central vacuole and a
nucleus, and much of the cytoplasm is occupied by a large
cup- or bell-shaped chloroplast or, more accurately, chromatophore. Near the base of each cilium is a small contractile
vacuole at one side of which is an orange pigmented spot
associated with a special receptor organ in the response of the
cell to light (including movement). This is obviously an
important factor in a photosynthetic and wholly motile
organism. All absorption of material for synthesis and all
exchange of respiratory gases must take place by diffusion
in solution across the cell wall and the protoplasmic surface.
Photosynthesis is probably concentrated round a special body
in the protoplast called the pyrenoid.

The cells usually reproduce frequently, the simplest method
being by straightforward division into two similar cells. In
other cases the parent cell loses or withdraws its cilia and the
contents divide into four to sixteen small individuals like
the parent, and called zoospores. They develop cilia and are
released when the old wall breaks down.

At other times the parental cell divides into more numerous
individuals which are without walls, and these are released as
before. This time, however, they unite in pairs, the members
of a pair normally coming from different parents. The
zygote so formed lays down a thick wall and becomes the
resistant zygospore which can withstand the complete drying
out of the pool in which the plants grew. Under favourable
conditions it divides to produce zoospores which grow into the
normal individuals. It is probable that a reduction division
occurs at this stage, so that the actual plant is haploid, i.e. it
contains one set of chromosomes only. Normally the gametes
are quite similar and are said to be **isogamous,** but in odd
species they differ at least in size, and are **anisogamous.**

The next stage in organisation is probably seen in colonial
forms. Such a condition is seen in Volvox. This plant consists of a hollow sphere formed of numerous individuals somewhat resembling Chlamydomonas with the cilia directed
outwards so that the whole sphere can be kept moving.
Within this " shell " of cells is a mucilaginous mass. However, not all the cells are alike: the vast majority are purely
vegetative and carry out the normal nutritional and respiratory
functions, but are incapable of reproduction, a condition which
seems to be one of the first restrictions on the cell in an
organised body. Among these cells are special individuals

which are concerned in reproduction. Of these, the partheno-gonidia are asexual and divide to form a flat plate of cells which ultimately rolls up to become a daughter colony inside the parent. Several of these may develop at once, and are released only when the parent structure disintegrates.

Other special cells develop into the female cells or oospheres and are rather larger individuals, whilst others again become the male organs or antheridia, and these in turn give rise to spermatozoids which are released to unite with the oospheres. A zygote is produced, and develops a thick wall becoming a zygospore or, in this case, having arisen from a definite oosphere, an *oospore*. In due course this germinates to produce a new colony.

This colonial development may represent the first stages in the formation of a thallus, particularly as the individual cells are linked by protoplasmic threads. Volvox frequently occurs in immense numbers in ditches, pools, etc., in the summer.

FIG. 55.—Part of filament of Spirogyra.

c.w. cell wall, *nu.* nucleus, *sp.ch.* spiral chromatophore, *py.* pyrenoid.

Another well known (though not always so common) form found in fresh water is Spirogyra (Fig. 55). This is a fila-mentous form which is really a longitudinal colony one cell thick. The cells are cylindrical, and are characterised by one or more spiral chloroplasts or chromatophores lying in the cytoplasmic layer lining the cell wall. Each cell is an in-dividual, and when fully grown divides independently, so that there is no specific growing apex, and if fragmented the bits of filament can continue to grow in isolation. Along each chloroplast lies a chain of pyrenoids, and there is a centrally placed nucleus. As in Volvox and Chlamydomonas, all absorption, etc., must take place by diffusion across the cell wall and cytoplasmic membrane.

There is no well-defined asexual reproduction, but a form of sexual reproduction known as conjugation can be observed. In this process each cell of the filament may become a game-tangium and its contents a gamete. Conjugation takes place between filaments lying alongside one another, and though there is no visible difference between the filaments, all the gametes of one act as males and those of the other as females. The actual transfer of the male gametes is done through special outgrowths from each filament, called conjugation tubes. As before, the conjugation produces a zygospore, which is released by the disintegration of the parental wall. In this plant reproduction means that part of the original plant is lost, since with the discharge of the male gametes the old male filament is just a series of empty cellulose walls. Conjugation may take place between filaments of very different cell size.

In other filamentous forms the structure is more advanced. Thus in Vaucheria, a coarse cylindrical form found on damp earth as well as in water there are no cross walls to divide the filament into separate cells, and the tubular thallus has many small nuclei distributed along the cytoplasm. Growth now takes place from one end of the thallus, the apex, and the older part ceases growth. At the apex the protoplasm is dense, but in the older regions vacuoles appear. The chloroplasts are small and discoid, being more like those of flowering plants. One special feature is that the plant appears to produce oil, and not starch.

Asexually it reproduces by **multinucleate** zoospores formed at the ends of branches, which thus become zoosporangia.

Sexual reproduction involves the formation of special organs either on the main filament or on special side branches. The important thing is that they are separated by a wall from the main vegetative structure and do not interfere with the normal activities of the vegetative body. The male organ is called the **antheridium,** and produces biciliate spermatozoids, whilst the female organ is the **oogonium**, and eventually contains a single large **oosphere** which is fertilised *in situ* by a spermatozoid. The zygote develops into an oospore with a thick wall and has a resting stage. Ultimately it germinates, and it is interesting to note that the young plant produces a colourless attaching thread or rhizoid (which later degenerates) as well as the normal green tube, thus showing a certain rudimentary differentiation. A thallus of the type found in Vaucheria, multinucleate but without separate cells, is called a coenocyte.

There are many other Green Algae, freshwater, marine and semi-terrestrial, and they show a great variety of form. Thus the Sea Lettuce (*Ulva lactuca*) is found on many shores and grows as thin flat sheets, whilst other green Algae appear as densely branching masses.

A more complex organisation is found in the Brown Algae. These are the familiar brown seaweeds found between the tide levels and in the deeper zones off our beaches. In this group the plant structure may range from small filamentous types to large thalloid plants, of which one form is estimated to reach a length of 200 feet. Most of them are attached to rocks or to the sand, but some are free floating, of which the most famous example is the Sargasso weed.

In these Algae the chlorophyll is masked by the brownish pigment phaeophaein or fucoxanthin. Sometimes the latter breaks down, and the chlorophyll is then evident.

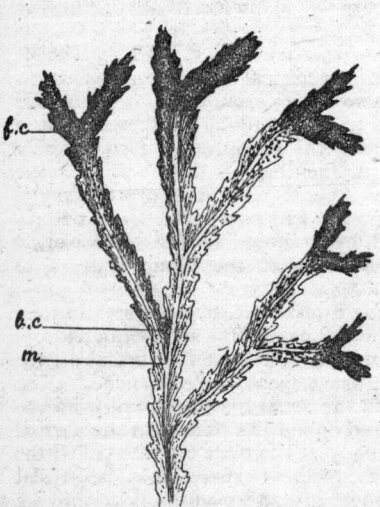

One of the commonest types of Brown Alga is Fucus, of which several species are found round our coasts occupying a fairly well-defined zone in the intertidal region. Two of the best known are *Fucus serratus* (the Serrated Wrack) and *Fucus vesiculosus* (the Bladder Wrack).

The mature plant is attached to the rock by a more or less irregular holdfast from which arises a cylindrical stipe. The upper part of the plant is a flat thallus which has a characteristic forked appearance due to the presence of an apical cell which by its divisions produces two " wings " of tissue. This is called dichotomous branching. Along the middle of the thallus is a thickened region called the midrib, and the stipe is actually the midrib in an older region from which the thinner " wing " region has been worn away. In a very young plant there is no stipe. Fig. 56 shows a frond of Serrated Wrack.

Fig. 56.—Part of Frond of *Fucus Serratus* (Serrated Wrack).

b.c. barren conceptacles, *f.c.* fertile conceptacles, *m.* midrib.

Internally the cells have cellulose walls, and a great deal of mucilage is present. At the surface of the thallus the layers of cells are packed tightly together, giving a regular parenchyma-like appearance, which is most marked in the outermost layer. This is called the outer limiting layer, and is the main photosynthetic region and also the meristematic layer. Most of the plastids are found in this layer.

Below the outer layer are several rows of cells less densely packed, and these are mainly storage cells containing large quantities of a carbohydrate called laminarin. The central region of the thallus shows a very loose arrangement of cells, and the mucilaginous walls are very swollen. This is the medullary region, and in the midrib part the long axis of the cells is parallel to the direction of the midrib, but they turn outwards in the wing region. Scattered over the thallus are cavities called conceptacles in which numerous branching hairs develop and secrete mucilage. In the Bladder Wrack there are also cavities filled with air which act as floats. Each conceptacle has a small opening or ostiole, and at low tide the mucilage escapes through this on to the surface of the plant and helps to prevent desiccation. The thallus is quite unprotected by anything like cuticle, and indeed the whole surface of the thallus is again the absorbing region.

Reproduction is mainly sexual. The organs are developed in fertile conceptacles (Fig. 57) (the general vegetative ones being referred to as barren conceptacles) towards the tips of the thallus, and the latter become swollen. In some species (*spiralis*) the male and female organs are developed in the same conceptacles, but in Serrated and Bladder Wrack the male and female plants are separate (dioecious). The male organs are antheridia and develop on branching hairs among which may be sterile hairs or paraphyses. Each antheridium has an outer wall and a thin inner wall and gives rise to a number of spermatozoids which have a pair of lateral unequal cilia and a bright orange plastid. When the antheridia are ripe the outer wall breaks down and the spermatozoids, still enclosed in the inner wall, escape out of the ostiole in the mucilage secreted at low tide. When the plant is covered by water again the wall enclosing the spermatozoids breaks down and they are released into the water.

The female organs are oogonia. Each has a short stalk and contains from two to eight oospheres, according to the species. The oospheres again are surrounded by a double wall. Paraphyses are present and secrete mucilage, so that when the oospheres are mature they are extruded like the antheridia. Again the inner wall breaks down when the tide covers the plant and the oospheres are released. Fertilisation thus takes place outside the plant, as we have the rather rare condition of the female cell being released. The oospore produced germinates at once and grows rapidly to form a small pear-

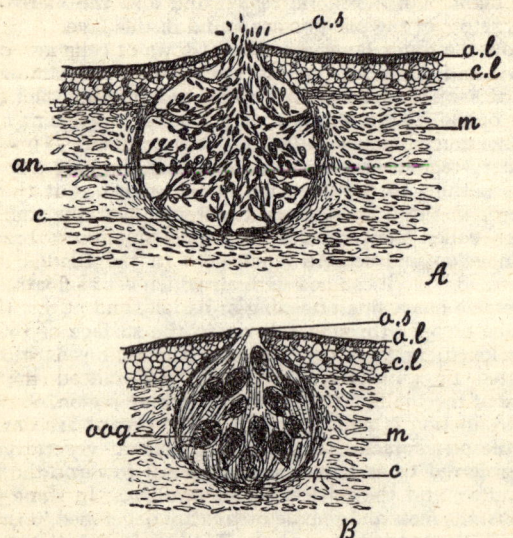

Fig. 57.—A. Section through Male Conceptacle of *Fucus serratus*.
B. Section through Female Conceptacle of *Fucus serratus*.

c.l. cortical layer, *c.* conceptacle, *m.* medulla, *o.l.* outer limiting layer, *os.* ostiole, *an.* antheridium, *oog.* oogonium.

shaped plant which anchors itself in a very short time and grows into the typical Fucus plant.

The Fucus plant is diploid, and is rather unusual among plants, in that the reduction division immediately precedes the formation of the gametes.

The oospore grows directly into a new plant similar to the parent, but in some Brown Algae there is an alternation of the sexual plant with an asexual plant—a condition known as an Alternation of Generations, which will be seen very definitely in some of the higher groups.

There are many forms of brown seaweed, and some of them are more complex than Fucus, but the latter gives a fair idea of the general organisation.

In the Red Algae there is still further complication in their life-histories, but their general organisation is very definitely still algal.

(2) The Fungi

In the Algae there was a gradual evolution of vegetative structure and an advance in reproductive methods.

In the group now to be considered there will be found much variation in reproductive methods, though it is doubtful whether these are advanced, whilst the vegetative body is always simple.

The Fungi have several features which can be regarded as characteristic. The plant body consists of filamentous strands which are known as **hyphae** individually, and together constitute the mycelium. The hyphae may be divided by cross walls, in which cases they are said to be septate, though the individual segments are rarely cells in the usual sense of being uninucleate units, or they may have no cross-walls, and are then said to be coenocytic. In some Fungi the hypha is a single cell, a condition which is most easily seen in yeast. In none of them is the vegetative body organised beyond the filamentous condition. There is no thallus more complex than this, and even the large structure found in such plants as the Mushroom consists only of interwoven hyphae closely aggregated and sometimes connected by transverse junctions. In the Slime Fungi or Myxomycetes the plant body consists of a multinucleate mass of protoplasm without cell walls.

The fungal cell wall is peculiar, in that it is not of the ordinary cellulose structure, but is a special material called fungal cellulose, which has nitrogen in it. In particular this permits the parasitic fungi to secrete enzymes which can attack the cellulose walls of the plants used as hosts without damage to their own walls.

One of the most conspicuous features of fungal organisation is the absence of chlorophyll, and because of this they are incapable of photosynthesis. Therefore they require synthesised carbohydrates for absorption, and these are obtained either from other living organisms (parasitism) or from the dead bodies of these organisms and from materials produced by them (saprophytism). All Fungi are parasites or saprophytes in some degree, though the association with other plants is not always clearly defined. Parasites may be a serious threat to other plants, whilst some saprophytes are destructive to foodstuffs, etc., which have been prepared for man's use. Although chlorophyll is not present, many of the

Fungi are pigmented. The Fungi do not form starch, but many of them produce glycogen, which is an allied substance similar to that found in animals.

One aspect of fungal nutrition is that some of them at least are able to use ammonium compounds as a source of nitrogen—an important advantage in acid soils.

Although the nutritional habits of Fungi may be destructive in the case of the parasitic types, their activities in the soil are very important in the carbon and nitrogen cycles. Moreover, many of the higher Fungi form relationships with the roots of Flowering Plants which enables an exchange of nutrients to take place. This is the mycorrhizal association. It is very important in acid soils, where nitrate formation is slow or difficult, and many moorland and forest plants are dependent on mycorrhizal relationships for the greater part of their nitrogen.

The Fungi show both asexual and sexual reproduction, but the latter has not been observed in all groups and is doubtful in others. Asexual reproduction is usually by small spores, uninucleate in some cases and multinucleate in others, and produced either in a sporangium or by constriction from the end of a fertile branch. In the latter case they are called conidia.

The Fungi are widely distributed and grow in a variety of environments, but always they need much moisture. Many of the lower Fungi will not grow if the relative humidity is below 80%, whilst even the larger ones are rapidly desiccated.

Three major divisions of the Fungi are recognised, based partly on vegetative features and partly on reproductive methods. The three groups are:

1. **Phycomycetes,** the most primitive group. They have no cross walls (septae) in the vegetative hyphae, and include two main sections, Oomycetes and Zygomycetes. Pythium and Cystopus, which are plant parasites, are examples of Oomycetes, and Mucor, the Pin Mould or Bread Mould, is a Zygomycete.

2. **Ascomycetes.** This is a very large group in which the hyphae are septate and the spores are produced in structures called *asci*, the formation of which is characteristically preceded by sexual reproduction. The group includes moulds such as Penicillium and Aspergillus, the parasitic types Erysiphe (e.g. Gooseberry Mildew) and Claviceps (Ergot), and some larger forms more like the gill Fungi (Toadstools). It also includes the unicellular Yeasts.

3. **Basidiomycetes.** This is the group which includes the forms familiarly known as Mushrooms and Toadstools, Puffballs and also the parasitic Rust Fungi. They have septate hyphae and produce spores on special organs called basidia which are arranged in various ways. Some of the Rust Fungi, etc., show an alternation of hosts and a complex life-cycle.

In addition, there is a group known as the Fungi Imperfecti. These are species in which the life-history is incomplete (or incompletely known) and which may be stages of species which could be properly included in the Ascomycetes.

A few examples will be chosen to illustrate the feature of members of these groups, but it must be remembered that there is much variation, and examples cannot be entirely representative.

1. **Phycomycetes.** The division of this group into Oomycetes and Zygomycetes is based mainly on reproductive differences. As an example of an Oomycete the common fungus *Cystopus candidus* is chosen.

This species is parasitic on the flowering plant Shepherd's Purse, on which the parasite can be seen as white patches on the stem and leaves and causing much distortion of the tissues. The nonseptate hyphae invade the air spaces of the tissues and penetrate the cells by means of structures called **haustoria**. These grow into the cells, and the contents of the latter are broken down and absorbed. In an allied form, Perenospora, the haustoria branch and fill the host cell. After a time the vegetative hyphae of Cystopus begin to produce asexual reproductive structures. The hyphae mass together below the epidermis, usually in the stem, and from the tip of each hypha a series of rounded bodies or conidia are formed by successive constrictions. Each is unicellular. The term conidia is usually retained for spore-like bodies produced by constriction from the tip of a hypha.

From each hypha a long chain of conidia may develop and the pressure ultimately breaks the epidermis so that the conidia are exposed. They now break off, and as they are very light they can be blown about to reinfect other hosts. The conidia are multinucleate, and as they are produced in large quantities, infection of the host species rapidly spreads. In some cases the conidia may give rise to ciliated bodies or zoospores. In either case the germination of the reproductive body produces a hypha which enters the tissue of the new host, probably through the stomata.

This is probably the most active method of reproduction, but Cystopus also produces sexual organs. These are found inside the host tissues and are developed in the intercellular spaces. The female organ is an oogonium and is formed at the end of a hypha by the laying down of a cross wall. The oogonium contains several nuclei at first, but when mature there is a central oosphere with one nucleus. Below the oogonium or from a neighbouring hypha a curved antheridium is also separated by a cross wall and produces a uninucleate male gamete (not a spermatozoid). The antheridium grows up against the oogonium and the male gamete penetrates the oosphere. The nuclei associate and a zygote is formed which develops into a resting oospore with a tough wall. The oospore is liberated only when the host tissue dies. When the oospore germinates reduction division occurs and a number of zoospores are formed which reinfect new Shepherd's Purse plants.

This cycle illustrates a very common feature—a rapidly multiplying asexual phase enabling wide distribution of infective spores and a sexual phase which provides a resting resistant structure.

Other Oomycetes show variations of this scheme, but there is much in common.

A good deal has been made of the similarity between the reproductive processes of some of these Fungi and those of certain Algae like Vaucheria in discussing the origin of the Fungi and their relationships with the Algae. Other points of similarity can also be used in working out this problem.

In the Zygomycetes a different type of cycle is seen. Many of the Zygomycetes are saprophytic and are found growing in decaying fruit or other materials with a ready supply of nutritive substances, e.g. bread and preserves.

One of the commonest of such species is the white mould Mucor found often on damp bread and other flour materials. It is often the first of a series of moulds which attack bread. The hyphae are non-septate and rather coarse, forming a dense tangle of mycelium. Some of the hyphae are actively engaged in absorption from the nutrient medium whilst others are on the surface. There is little difference between them and the surface hyphae would very rapidly become absorbing ones, but while they are exposed their walls are probably rather thicker.

Because the fungus requires a good supply of oxygen, the hyphae do not penetrate deeply into a substrate unless it is well aerated.

Growth occurs from the tips of the hyphae, which branch

repeatedly. If the fungus is inoculated on to the surface of a clear nutrient medium like agar or gelatine, the growth of the hyphae can be observed easily with a microscope. It may be noted that the growth radiates outwards from the point of inoculation, and indeed the central part of a big colony may die back as the available food is exhausted. Mucor produces oil globules in the cytoplasm.

After a very short time hyphal branches grow vertically into the air and the tips swell out. In each tip a mass of protoplasm with many nuclei is formed and a cell wall is put down to separate the tip from the stalk. This terminal region is the sporangium and the stalk is the sporangiophore. The protoplasmic mass divides into multinucleate spores, and these are pressed against the sporangium wall by the continued growth

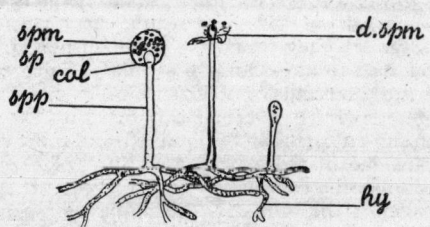

FIG. 58.—Sporangia of *Mucor*.

col. columella, *d.spm.* discharged sporangium, *hy.* hypha, *sp.* spores, *spp.* sporangiophore, *spm.* sporangium.

of the top of the sporangiophore, which forms a club-shaped columella within the sporangium (Fig. 58). The wall becomes thicker and darker and is rather rough. Ultimately the pressure within (aided perhaps by the shrinking of the wall due to drying) bursts the sporangium and the spores are shed to reinfect a suitable medium. Immense numbers of spores are produced, and if a nutrient plate is exposed to the atmosphere it will be very unusual not to find fungi developing within twenty-four hours of exposure.

In Mucor the sporangia are single, but in the allied species Rhizopus several sporangia arise at one point. The asexual method is the general type of reproduction, but a form of sexual reproduction known as conjugation can be observed under appropriate conditions This method of reproduction is very spasmodic under natural conditions, but can be observed regularly in the laboratory by simple culture methods. When

it occurs short lateral hyphae grow out from adjacent filaments and eventually they meet end to end. In the meantime much protoplasm and many nuclei have appeared in the
terminal portion of each of the lateral hyphae, and this region
is now separated by a cell wall. Thus each hypha has produced a **gametangium** containing a multinucleate gamete.
After the gametangia have made contact the dividing wall
breaks down and the two multinucleate gametes conjugate,
the respective nuclei fusing in pairs. The resulting zygote
develops a thick wall which becomes roughened and black, and
this zygospore may remain dormant for some time after separation from the hyphae on which it developed. Eventually it
will germinate and produce a single sporangium only, from
which the spores are shed and germinate in the usual way.
Mainly due to the work of Blakeslee, it was demonstrated that
conjugation occurs only when two different strains of Mucor
are present, though they cannot be distinguished structurally.
They are referred to as + and − strains. Cultures of these
are kept (as are thousands of other species) in such institutions
as the Commonwealth Mycological Institute, from which
organisation sub-cultures can be obtained, and if the two strains
of Mucor are inoculated on to a culture plate (especially
gelatine), conjugation can readily be observed. It is probable
that the nuclei of the ordinary Mucor plant are haploid and
those of the zygospore diploid, reduction division taking place
when the spores are formed in the first sporangium. Each
spore is either + or − and thus the two strains appear again.
Rhizopus is very similar to Mucor, but the sporangia are much
larger, the hyphae coarser and a group of sporangia usually
arise at one point.

2. Ascomycetes. This group includes many filamentous
microfungi, but also some species which produce large
fructifications. The characteristic features of the group is
that the plants have hyphae which are divided by cross walls,
i.e. septate. The segments thus produced are not really
comparable to an ordinary plant cell, as they frequently have
more than one nucleus. There is a sexual reproductive
process which typically involves an **ascogonium** and **antheridium** and leads to the production of special bodies called
asci which contain the **ascospores.** The actual procedure
varies very considerably, and in many cases it seems likely that
there is no actual conjugation even though both organs may be
formed. Asexual reproduction is frequent and is brought
about by the formation of conidia at the apices of special
hyphae.

Examples of Ascomycetes are the common moulds Penicillium and Aspergillus, the various Yeasts (which with the moulds are saprophytic examples), Sphaerotheca (Gooseberry mildew), Erysiphe and Claviceps (Ergot), which are parasitic types. Other examples include the Toadstool-like Morel and some of the cup-like Discomycetes.

Various species of Penicillium and Aspergillus appear on fruit, bread, etc. very often succeeding Mucor. The mycelium consists of very fine whitish hyphae which rapidly cover the substrate. It is probable that they will tolerate somewhat drier conditions than Mucor and they certainly seem to persist much longer on an old piece of bread.

After a very short period the septate hyphae begin to produce conidiophores—upright branches which give rise to the conidia. In Penicillium the conidia are produced on finger-like branches with varying degrees of division (Fig. 59). The ultimate parts of the branches are called sterigmata, and these produce a rapid succession of conidia. An excellent culture can be obtained from one of the common " blue " cheeses. Aspergillus always forms club-shaped conidiophores, from which the sterigmata grow, and again the conidia are constricted from the ends. The conidia are uninucleate and vary in colour according to the species. They are produced in such dense masses that the whole fungal growth appears coloured, so that while *Penicillium notatum* appears bluish-green, *P. luteum* is yellow, whilst in the case of Aspergillus the species *niger* is black and *A. oryzae* is yellowish. In normal circumstances the chains of spores are rapidly broken and dispersed by air currents and the spores soon germinate on a suitable medium.

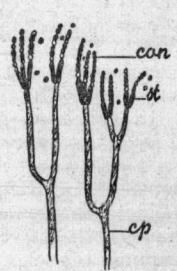

FIG. 59.—Conidia of *Penicillium*.

cp. conidiophore, *con.* conidia, *st.* sterigma.

Sexual reproduction is frequently found in some species of Aspergillus, especially when the culture is getting old. In this process certain hyphae give rise to the reproductive organs, the latter usually being in close proximity. The female structure is usually terminal and the male lateral, though both may be end cells. Contact is made through a region known as the trichogyne, and the male gametes pass into the ascogonium, where the fertilisation of the female gamete is completed. The degree to which this process is established in the group is very variable, and it is suggested that in many Ascomycetes further development of the female structure may occur without fertilisation, the male structure simply dying away. Even so there is always a nuclear fusion before ascospore formation.

The next stage is quite general. The fertilised nucleus divides and the cell produces branches which are called **ascogenous hyphae,** into which the nuclear products pass. Here they undergo further division and part of the hypha becomes an ascus with eight ascospores. In the meantime a number of hyphae have grown out from below the ascogonium and by branching and coiling have enclosed it. They actually grow so tightly together that they form a wall which is the perithecium, and it is in this that the asci are formed. As it ripens the wall becomes hard and often dark coloured, whilst the actual asci may break down so that the ascospores lie free in the perithecium. In some forms, such as Erysiphe, Chaetomium and others, the perithecia may bear long spike-like outgrowths and are big enough to be seen by the naked eye.

In *Claviceps purpurea* (Ergot) the hyphae grow in the developing ovary of grasses such as Rye or Wheat and their wild relatives. When the ear is ripe some of the grains are replaced by hard black bodies known as **sclerotia.** These fall to the ground, where they pass the winter (unless they have been harvested with the grain). In spring or early summer the sclerotium sends out a number of structures which are really stalked perithecia. In these the asci develop, and the spores can infect a new crop of grasses. It seems likely that the sclerotia do not germinate unless they have been subjected to frost.

The Yeasts or Saccharomyces have several interesting features. Firstly they are probably the fungi which have been used longest by man because of their action in fermenting sugars to produce alcohol, and secondly they are unusual in being unicellular.

The common brewers' Yeast, *Saccharomyces cerevisiae*, has ovoid cells in which the nucleus is at the end of a large vacuole, across which are strands of cytoplasmic material. It is doubtful whether the young Yeast cell has a true cell wall, but this is certainly present in the older individuals. In the cytoplasm are granules of glycogen and another reserve, volutin. Yeasts are found occurring naturally on the surface of fruits, etc., and obtain the substances necessary for growth by absorption through the surface of the cytoplasm. They obtain energy by the fermentation of sugars, and when this process is carried out in the presence of an insufficient supply of oxygen it results in the formation of ethyl alcohol, to which the Yeast is relatively tolerant. This is the basis of brewing practice, the formation of large amounts of carbon dioxide helping to exclude oxygen. This anaerobic breakdown is largely due to the action of the enzyme zymase, and it will be recalled that the study of zymase action has played a large part in research on respiration. By changing the conditions under which the Yeast acts it is possible to produce other by-products of fermentation. Before the facts of the process were fully known it was customary in wine-making to allow the wild Yeasts on the surface of the grapes to bring about the fermentation. Pasteur showed how this could be affected by other Yeasts which were, to say the least, less useful, and it is now the practice in all controlled fermentations to use specific strains of Yeast introduced under strict conditions.

During the fermentation the Yeast reproduces by a process known as budding. The nucleus divides and a small protrusion appears on the side of the cell. One of the daughter nuclei migrates into this, and the bud or **gemma** increases in size and finally separates (functionally if not for some time physically) from the parent cell. Before this happens the original cell may have produced more buds and the daughter cells themselves may have budded. The process proceeds faster in a well-aerated solution, but eventually slows down and all the cells separate. Under these conditions the Yeast reproduces very rapidly and is in fact a by-product of some brewing processes.

If the Yeast becomes exposed to adverse conditions such as dryness or shortage of suitable nutrients the cell may lay down a tougher wall and inside this the protoplasm divides into two, four or eight daughter cells. The original cell has become an ascus and the daughter cells are ascospores. It has now been established that the procedure may be associated at some stage with a conjugation process, and it has been possible to hybridise Yeasts to produce particular strains for specific purposes.

3. Basidiomycetes. In this group the hyphae are again septate and the " cells " are frequently binucleate. The mycelium may be parasitic or saprophytic, and not infrequently both. In saprophytic types the mycelium is usually found in the soil, and dense masses may be seen by turning over the " leaf-mould " in a wood. Many of the Basidiomycetes are filamentous at all stages, and such cases are seen in the Rust Fungi and the Smuts which attack cereals and other Flowering Plants. Other Basidiomycetes produce massive spore-bearing bodies, and these include the gill Fungi, the typical Mushrooms and Toadstools, the Bracket Fungi seen on trees, and other forms. Nevertheless it must be emphasised that there is no tissue organisation more complex than the filamentous hypha.

Reproduction in this group is by basidiospores. They are developed on special cells called basidia and their actual production is similar to that of conidia; that is to say, they are constricted from the ends of sterigmata. The main difference is that such development is limited, each basidium producing a single group of spores only.

Normally four basidiospores are formed, but this number may vary—in the cultivated Mushroom, for instance, there are only two. In the majority of cases there is little evidence of a sexual process, though nuclear exchange and fusion may take place in the hyphae below the spore-bearing layer.

The Rust Fungi can be very serious plant parasites and may produce epidemic disease in cereal crops. This has been controlled in various ways, but particularly by breeding rust-resistant forms of the plants.

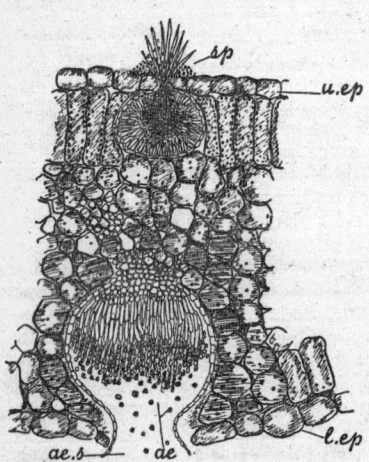

FIG. 60.—Vertical Section of Barberry Leaf with Aecidiospores and Spermogonium.

ae. aecidiospores, *ae.s.* aecidiosorus, *sp.* spermogonium, *u.ep.* upper epidermis, *l.ep.* lower epidermis.

Puccinia graminis, the Black Rust of Wheat, is a species in which the life-cycle has been fully worked out. Although one may describe the life-cycle of the " species ", there are many races which are restricted to different varieties of Wheat or other cereals, and these are known as physiologic races. They present a great problem in the development of rust-resistant varieties, but much has been done to cut out the ravages of these plant parasites. The main feature of this fungus is that it requires two hosts. In spring orange spots or sori appear on the underside of leaves of the common Barberry. From these sori are produced the first kind of spore—the aecidiospore (Fig. 60). The fungal hyphae have grown among the cells of the Barberry leaf and finally massed to produce the sorus or cluster-cup, the development of which causes a break in the leaf-surface. From the ends of the hyphae forming the sorus a succession of binucleate aecidiospores is produced, the spores being separated by small intercalary cells. These aecidiospores cannot reinfect the Barberry, but if they reach a Wheat plant they germinate, invade the tissues through the stomata and parasitise the cells of the Wheat plant. During the summer the hyphae

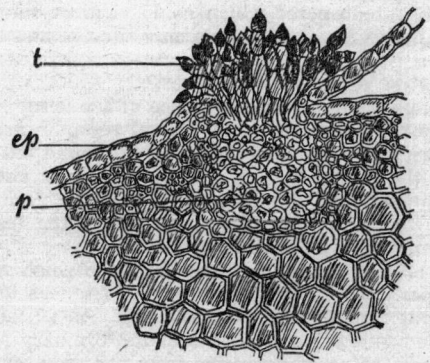

FIG. 61.—Teleutospores of *Puccinia graminis* on Wheat Stem.
ep. epidermis, *p.* parenchyma, *t.* teleutospores.

mass below the epidermis of the leaf (in particular) and produce another type of sorus known as the uredosorus. In this structure single-celled stalked spores are formed, each being binucleate. These uredospores or summer spores can infect more Wheat plants and are responsible for the epidemic phase of the disease. New sori are produced throughout the summer, but as the plant ripens the same sori, as well as new ones on the stems, give rise to a third type of spore called a teleutospore or winter spore. These again are stalked, but have two binucleate cells. Teleutospores are illustrated in Fig. 61. The walls are thick and black, and these spores are resistant, lying on the ground or on the straw throughout the winter. In spring they germinate and each cell gives rise to a four-celled basidium, and from each cell a spore is produced. It is this spore which reinfects the Barberry.

The cycle is a complex one, and different species of Puccinia have different hosts, whilst in some cases the cycle may be completed on one host. In some cases the complete life-cycle is not known, and whilst one or more stages may be familiar, others are either not known or have not been recognised as part of the sequence in that particular fungus. If they are actually absent it means that there are some Rust Fungi which do not depend on the complete series of forms. Control of this disease was first attempted by destruction of the Barberry plant. The basidiospores cannot affect the Wheat plant, and in temperate climates the winter is too severe to permit the survival of the uredospores which could reinfect the crop. In warmer climates the uredospores may survive and be responsible for the persistence of the disease—a source of trouble in some potential Wheat-growing areas.

In Puccinia there is some evidence of a nuclear association which may represent a form of sexual reproduction. Structures known as spermogonia (Fig. 60) are found on the upper surface of the Barberry leaf, and they produce two kinds of cells which fuse, and it is after this that the aecidiospore is produced. Such a procedure has not been established, however, in many of the other species.

The gill-bearing Fungi in general are those which produce the largest types. The spores are produced from basidia, and in the true gill Fungi these basidia cover the surfaces of flat plates called gills or lamellae, as in the ordinary Mushroom. In such cases these gills radiate from the stem or stipe and lie below the cap or pileus. In the Bracket Fungi and some toadstool-like forms the basidia are found lining tubes on the under-surface of the cap, whilst in other forms again they are on long tooth-like projections. In the Puffball group of Fungi the spores are produced on a mass of hyphae called the gleba within the more or less globular body called the peridium. When the spore-bearing structure is ripe the peridium bursts either to release the spores directly or to produce a special organ on which the spores are borne.

The gill-bearing Fungi are probably among the most familiar types because they are large and often brightly coloured and also because some of them are edible.

The common Mushroom, *Psalliota campestris*, is one of the best-known types either in its wild form or as the cultivated species which has only two spores on the basidium. The vegetative part of the fungus consists of fine whitish hyphae

which are septate and in general binucleate. The hyphae are
saprophytic and are found where there is plenty of organic
material in the soil. It is the hyphae which are sold in blocks
mixed with a suitable compost and called " spawn ".

When reproduction occurs the hyphae mass together and
form small button-like structures of interwoven hyphae.
They enlarge and gradually differentiate into several regions.
The upper part, which is the future cap or pileus, is supported
on the stalk or stipe. Under the cap the gills or lamellae
begin to differentiate, and at first they are enclosed in a

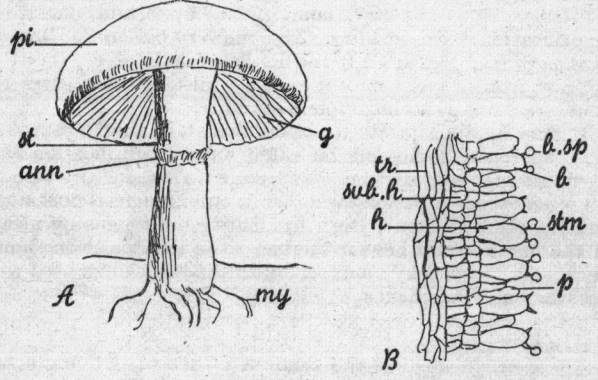

FIG. 62.—A Mushroom (*Psalliota* sp.)

A. General Appearance. B. Detail of structure of part of Gill.

ann. annulus, *b.* basidium, *b.sp.* basidiospore, *g.* gills, *h.* hymenium,
my. mycelium, *p.* paraphysis, *pi.* pileus, *st.* stipe, *stm.* sterigma,
sub.h. sub-hymenium, *tr.* trama.

chamber by a veil of tissue called the velum, stretching from
the stipe to the edge of the pileus. As the Mushroom grows
the cap expands and the velum splits, leaving a ring or annulus
around the stipe (Fig. 62). The gills are now exposed and
radiate from the stipe, though not all of them extend from the
stipe to the edge of the pileus. In vertical section each gill is
V-shaped, and in such a section the hyphae can be seen to turn
outwards from a central mass called the trama to produce a
regular layer called the subhymenium, which in turn gives rise
to the actual sporing layer or hymenium. Here large basidia
are formed, and between them are sterile cells or paraphyses.
The basidia produce four (or in a few species two) horn-like

sterigmata, and from these the purple, binucleate spores are eventually discharged. If the cap of a Mushroom is placed on a sheet of white paper a pattern of the gill arrangement will quickly be produced by the discharged spores, of which millions are formed in a single fructification. Although it is not possible to distinguish a sexual process morphologically, the formation of basidiospores is always preceded by a nuclear fusion.

The method of basidial arrangement differs somewhat in the various groups, and the gills may be replaced by pores, by teeth or by a loose framework within an outer covering (the Puffballs, etc.). Spore colour is an important factor in identification. In some forms, like the very poisonous *Amanita phalloides* (the Death Cap) and its less deadly allies, the stipe appears from a basal cup or volva, whilst in other cases the velum remains as an incomplete curtain.

It may be appropriate to say at this stage that Mushrooms or Toadstools should not be eaten unless identification as a harmless species is certain. There are no satisfactory rule-of-thumb methods for telling whether or not a fungus is poisonous, and the student should learn to identify the fungus by means of the many useful keys available. The identification should always be checked by an expert until the key can be used with accuracy and confidence.

Bacteria

These organisms are the smallest of accepted living structures, though this term may have to be applied to Viruses and Bacteriophages. Superficially the structure of Bacteria is very simple, and consists of cytoplasm with a chromatin unit which may be a nucleus. The cytoplasm is not enclosed in a wall, but there is a protective membrane and often a capsule of some mucilaginous material. The bacterial cells are of several forms and are extremely small, a large rod type being about 5/1000 mm. long by about 1/1000 mm. in diameter. The cells may be spherical (cocci), rod-shaped (bacilli), comma-like (vibrio) and spiral (spirillum and spirochaete). They frequently possess fine protoplasmic outgrowths or flagellae which enable them to move about. A few of the rod-like forms produce very resistant endospores. Reproduction is normally by transverse fission, and under favourable conditions may occur once every twenty minutes. Bacteria have an enormous variety of nutritional and physiological activities, and there are few media which do not support some form of Bacteria. Some of them are parasitic and some are saprophytic, whilst many are facultative, which means that they can exist in either way. In spite of the fact that Bacteria are commonly regarded as being associated with disease, only a comparatively small number are actually pathogenic, the vast majority being relatively harmless.

There is an enormous population of Bacteria in the soil, where they are concerned in almost every type of breakdown and synthesis. The ultimate growth of many of the higher plants depends to a great degree on the changes brought about by Bacteria. Their wide range of physiological adaptability is almost certainly due to the variety of enzymes which they produce.

Many of the Bacteria, particularly the saprophytic ones, can be grown in culture in the laboratory, so that it is possible to make a close and detailed study of their behaviour. The use of the electron microscope has added much to our knowledge of their structure, but much still remains to be discovered about their activities, and particularly about their relationships with other organisms, especially in the soil.

Lichens

The Lichens are plants in which the thallus is a complex of Algal and Fungal cells. They are mainly small plants and are found on the ground, on tree trunks, stones, etc., and are often the first plants to colonise bare rock. The relationship between the two partners is rather complex, but it seems likely that the Fungus benefits from the photosynthetic activity of the Alga. The plant body is usually rather flattened and thalloid, but the reproductive structures may vary quite a lot. Most of the Fungi involved are Ascomycetes, so that the reproductive process includes the formation of Asci.

2. BRYOPHYTA—THE MOSSES AND LIVERWORTS

The plants in this group are all comparatively small and are dependent on a damp environment for successful growth, though many of the Mosses at least can survive considerable desiccation.

The plant-body is thalloid in many Liverworts, but in the foliose or leafy Liverworts there is a well-developed stem and leaves towards which transitional forms can be recognised. All the Mosses show the differentiation into stem and leaves. None of the Bryophyta possesses true roots, but there are colourless rhizoids which are unicellular in Liverworts and multicellular in Mosses. They attach the plant to the ground and are responsible for some absorption. There is no true vascular tissue, though in some Mosses there is a central region of elongated cells in the stem which help to support it, and it has been suggested that there may be some conduction along them. Nevertheless if the base of the plant is kept in water and the top exposed, the upper part soon shrivels, suggesting that the conduction is not very efficient.

The thallus or leaves (as the case may be) have discoid chloroplasts and are usually very thin. In the Mosses the leaves have a central midrib of elongated cells, and in some cases there may be rows of vertical lamellae acting as additional photosynthetic tissue. In some thalloid Liverworts there are chambers perforated on the upper surface by a simple pore, and

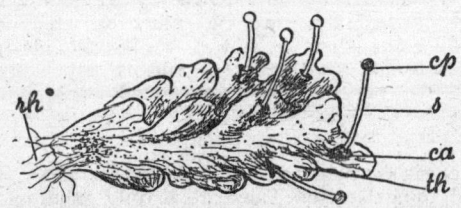

FIG. 63.—Portion of Thallus of *Pellia* with Sporogonia.
ca. calyptra, *rh.* rhizoids, *s.* seta, *th.* thallus, *cp.* capsule.

in these photosynthetic filaments are found. Most of the basal tissue is then colourless, and may act as storage cells. In both Mosses and Liverworts there is little differentiation from the simple parenchymatous type of cell.

In the thalloid Liverworts the structure is fairly simple (Fig. 63). The thallus is flat with a thickened central region, and at the apex of each lobe is a two-sided apical cell the division of which produces a more or less V-shaped growth which is referred to as **dichotomous** branching. Usually the older parts of the thallus die away. Below the thickened central region is the greatest concentration of rhizoids.

There is a gradual progression of types in which a midrib region becomes more apparent and also a gradual lobing of the wing part of the thallus (*Metzgeria*, *Blasia*). The foliose Liverworts are all trailing and show some complexity of leaf-structure. The leaves are in two rows, giving a flat appearance to the shoot, and in most cases part of the leaf develops as a lobe on the under-side—the under-leaf which is very elaborate in the species *Frullania* and *Scapania*. Only in very rare cases do the leaves have a midrib. Rhizoids appear along the under-side of the stem.

In the Mosses many of the species grow upright (**acrocarpous**) (Fig. 64) and may be several inches tall, as in some species of *Polytrichum*, the Hair Moss. Others show the trailing habit (*Hypnum, Eurynchium*), and these are called **pleurocarpous**. The leaves are in three rows (with the exception of *Fissidens*), there is always a midrib (except in the Sphagnales) and there are no under-leaves. In the Bog Moss, *Sphagnum*, the leaves have two kinds of cells, narrow bands of photosynthetic tissue enclosing wide dead cells with bars across their walls. These leaves are spongy and can hold a great deal of water (and presumably air).

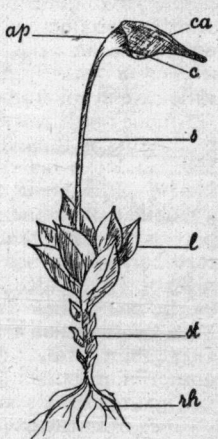

The Mosses branch freely—the upright ones from the base so that a tufted habit is common, whilst the trailing forms branch irregularly. In both cases there is a tendency to form a dense cushion of plants, often growing over a considerable amount of dead material.

The foliose Liverworts and Mosses occupy a variety of habitats, provided that the latter are sufficiently damp. They may be found on the ground, forming the lowest plant tier along with Flowering Plants, they frequently grow on tree bark and are often found on rocks, walls, etc. Some of them appear as early colonisers of bare patches of

FIG. 64.—Moss Plant with Sporogonium (*Funaria*).

ap. apophysis, *c.* capsule, *ca.* calyptra, *l.* leaves, *rh.* rhizoids, *st.* stem, *s.* seta.

ground, a particular example of this being *Funaria hygrometrica*. On the whole, Mosses will survive more arid conditions than Liverworts.

Both Mosses and Liverworts reproduce quite freely by vegetative methods—often by fragmentation. There is, however, a more specific method of vegetative propagation by means of small cell masses called **gemmae.** These are produced in a variety of situations; some occur in cups on the plant surface, as in the thalloid Liverworts Marchantia and Lunularia, whilst in the moss *Aulacomnium androgynum* they are formed in a ball at the end of an apical stalk and in the moss *Tetraphis pellucida* gemmae are produced in a cup-shaped cluster of leaves at the apex of the stem. In the latter case they may replace the sexual organs. In many plants, however, the gemmae are produced more or less haphazardly. Each gemma is a small mass of chlorophyll-bearing cells which eventually drops off the plant and germinates to produce a new plant.

The Bryophyta, however, show a well-organised system of sexual reproduction which involves the formation of very distinctive organs. The general structure of these organs is similar in Mosses and Liverworts, but they are rather more elaborate in the former.

Among the Liverworts the sexual process is observed very readily in the thalloid form Pellia. The organs develop about May, and the male structure is called an **antheridium**, the female organ being the **archegonium**. Each antheridium is a globular body and is sunk in a chamber in the median thick portion of the thallus, several antheridia occurring close together. Each has a wall of cells, and this wall encloses a mass of tissue which in due course gives rise by mitotic divisions to the coiled male gametes or **spermatozoids**. These bodies have two flagellae at the narrow end. The antheridium has a very short stalk. When the spermatozoids are ripe the antheridium bursts probably by the absorption of water, and the spermatozoids are released in a mass of mucilage which escapes from the cavity on to the surface of the thallus. It is essential that a film of water should be present in order to allow them to move about.

The archegonia are found at the growing apices of the thallus protected by a flap of tissue called the **involucre**. Each archegonium is club-shaped, with the swollen part attached to the thallus. The enlarged region is called the **venter**, and from it extends the *neck*. In Liverworts the whole organ is rather slender. The venter has a short stalk and its wall is only one cell thick, whilst the elongated neck consists of several (usually four or five) longitudinal rows of cubical cells enclosing a narrow neck-canal. Inside the venter is a single egg-cell or oosphere, and above this is a series of degenerate oospheres, of which the first is the ventral-canal cell and the succeeding ones the neck-canal cells. When the archegonium is ripe the neck-cells separate, possibly due to the swelling of the canal-cells, which now disappear. It seems likely that they produce a substance having a chemical attraction for the spermatozoids, and in Mosses this has been identified as cane-sugar. The spermatozoids enter the neck and one of them fertilises the oosphere. Development of the zygote or oospore begins at once with the formation of a cell wall followed by rapid division of the cell. As the cell mass enlarges it becomes differentiated into several distinctive regions. The surrounding tissue is also stimulated into further growth, so that for some time the venter of the archegonium keeps pace with the development of the embryo. Fig. 65A is a longitudinal section of the thallus of Pellia showing the sexual organs.

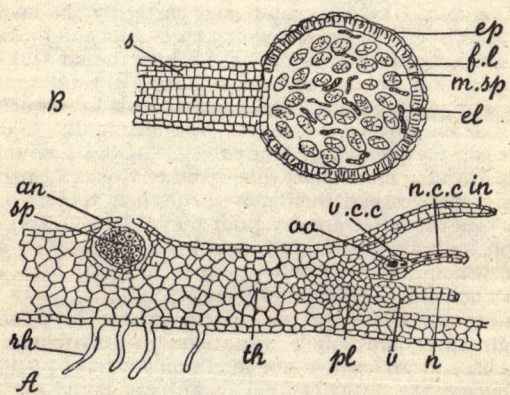

Fig. 65.—A. Longitudinal Section of Thallus of *Pellia* showing Reproductive Organs. B. Enlarged View of Capsule.

an. antheridium, *el.* elater, *ep.* epidermis, *f.l.* fibrous layer, *in.* involucre, *m.sp.* multicellular spore, *n.* neck, *n.c.c.* neck canal cell, *pl.* placenta, *oo.* oosphere, *rh.* rhizoid, *s.* seta, *sp.* spermatozoids, *th.* thallus, *v.* venter, *v.c.c.* ventral canal cell.

The growing zygote or embryo is called the **sporogonium** and is a new phase in the life-history; it must be regarded as a new plant. As a general rule only one sporogonium develops from each group of archegonia and it quickly shows three regions. At the base is a conical structure which penetrates into the parental tissue and acts as an absorbing organ. This is the **foot**. Above this is a region of regular rows of cells which are capable of rapid elongation, and these constitute the stalk. Finally at the free end there is formed the capsule, which is a spore-bearing structure in which a mass of sporogenous tissue is enclosed in a wall of cells. Within the wall the sporogenous tissue undergoes repeated division and gives rise to two types of cells. One type is the spore mother cell, and each of these divides by a reduction division to produce four cells which are the potential spores. Each spore then becomes multicellular and produces chlorophyll, and this condition may represent partial germination. The other type of cell elongates and the structures so produced are called **elaters,** each having a spiral band of thickening. These bodies are characteristic of Liverworts. In Pellia the sporogonium remains short during the winter with the capsule protected by the calyptra (the old venter) and partly by the involucre. About March or April the seta elongates rapidly and the capsule bursts the calyptra and is pushed up into the air (Fig. 65B). The inner wall of the capsule is irregularly thickened by fibrous strands, and in Pellia, as in most Liverworts, the opening of the capsule is achieved by splitting due to drying, the ripe capsule opening by four valves. At the same time the elaters with their spiral thickening tend to twist violently with humidity changes and their movements help to flick out the spores. As the latter are flung out they may be dispersed by wind or by water, and they germinate to produce a new thallus, which may in some cases arise as a bud from a rudimentary structure called a protonema.

This reproductive sequence is fairly general in Liverworts, although there are minor variations. In Marchantia the antheridia and archegonia are found on separate plants, and in each case the actual sexual organs are borne on special stalked structures. Fertilisation in such cases may have to be achieved by splash distribution of the spermatozoids, although the female structure does not elongate until after fertilisation. The sporogonia have capsules similar to those of Pellia, but without the long seta. In the foliose Liverworts the archegonia are found at the apices of the shoots, whilst the antheridia occur in the axils of the leaves.

In Mosses there is a similar form of reproduction, but on the whole the organs are rather more elaborate. The archegonia and antheridia are borne at the apices of the gametophyte plant. The sexes may be quite separate, as in the Hair Moss, *Polytrichum,* or they may be on the same plant with the archegonia on short side-shoots. Normally the antheridial plants (or shoots) are quite distinctive, as the apex often consists of a whorl of leaves enclosing a tightly packed mass of antheridia and sterile hairs or paraphyses (Fig. 66), the whole collection often having a slightly orange colour. This arrangement is called a **perichaetium.** Although the same

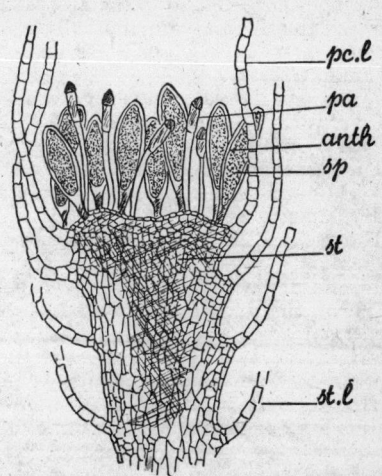

FIG. 66.—Vertical Section through Antheridial Head of Moss Plant (*Mnium*).

anth. antheridium, *pa.* paraphysis, *pc.l.* perichaetial leaf, *st.* stem, *sp.* spermatozoids, *st.l.* stem leaf.

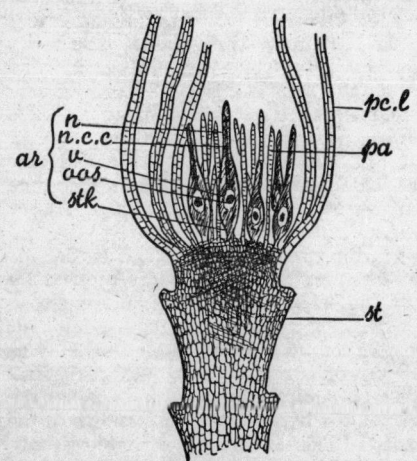

FIG. 67.—Vertical Section through Archegonial Head of Moss Plant (*Funaria*).

ar. archegonium, *pa.* paraphysis, *pc.l.* perichaetial leaf, *n.* neck, *n.c.c.* neck canal cell, *oos.* oosphere, *st.* stem, *stk.* stalk, *v.* venter.

arrangement is found in the archegonial heads they are much less distinctive and cannot usually be recognised by mere inspection, as they very much resemble an ordinary leafy shoot (Fig. 67).

The sexual organs themselves are more massive than those of Liverworts. The antheridia are elongated and club-shaped, with a definite stalk, and they occur in considerable numbers. They open in a manner similar to those of Pellia.

The archegonia probably represent the most elaborate development of this type of female structure in living plants. The neck is very long, the venter wall is several cells thick and there is a stout multicellular stalk. Fertilisation is again dependent on water for the transport of the spermatozoids. Where the archegonia are borne on short side-shoots the spermatozoids are probably washed on to them fairly easily, but in the tall mosses like Polytrichum, where the sexual organs are on separate plants, fertilisation is probably very irregular. It is interesting to note that in Polytrichum the stem apex continues to grow through the antheridial cup and a series of these may be formed.

Development of the sporogonium starts as in Liverworts, and again there is normally only one developed from each apex. A foot is pushed down into the gametophyte and a seta differentiated, but the whole structure is at first an elongated, rod-like mass carrying the remains of the venter. The lower region produces a foot and seta which are not easily distinguished at first, but as the seta elongates it turns green, whilst the foot is of course buried in the stem apex. The upper part of the embryo differentiates into a rather complex capsule. The base of the capsular region is called the apophysis and contains chlorophyllous tissue and has true stomata, so that it is capable of active photosynthesis. Above this there is a central strand of cells—the columella—around which the true spore-bearing tissue develops. Outside the sporogenous layer there is a region of air spaces crossed by bands of chlorophyll-bearing cells, and this is finally enclosed by the wall of the capsule. As the capsular region develops it becomes very clearly marked off from the seta and remains covered by the upper part of the old venter—the calyptra. At the apex of the capsule a conical cap is formed, called the operculum, and this is attached by a ring of cells known as the annulus. In the majority of Mosses the actual spore-containing region is closed by a layer of tissue called the peristome, which, because of the peculiarities of thickening, eventually separates into a number of teeth, of which there may be two rows. These are spirally arranged and have transverse thickened bands.

As the capsule develops the sporogenous tissue undergoes repeated divisions, resulting in the appearance of the spore mother cells. These undergo reduction division to give the actual spores. Much of the internal tissue breaks down so that the mature spores lie loosely in the capsule. In the meantime the calyptra falls off as it dries, whilst the annulus cells shrink and split, thus freeing the operculum. The peristome teeth open and close with changes of humidity and the spores are shaken out. There are no elaters.

In a few cases, as for example the Bog Moss Sphagnum, the spores lie in a region capping the columella and the capsule opens by a lid only.

When the spores are discharged they may be dispersed by wind, and eventually some of them reach an environment in which they can grow. The first structure produced is the protenema which is a filamentous branching organ with rhizoids. From this the typical Moss plant arises as a bud. Thus there is an intervening vegetative method of propagation because the protonema is a persistent structure. It should be pointed out that although the spores of Bryophyta have some degree of protection, they are not capable of prolonged resistance to adverse conditions such as desiccation, etc.

In both Liverworts and Mosses it is seen that there are two distinct phases in the life-cycle of the plant. The structure which we know familiarly as the Moss or Liverwort plant is the **gametophyte** and is haploid, that is to say it has the reduced chromosome number and its gametes are produced by direct division. As a result of fertilisation a new generation is produced consisting of the sporogonium, and this is diploid. In the Bryophyta this **sporophyte** generation is never independent, though in Mosses it is capable of a good deal of photosynthesis and probably receives only water and salts from the gametophyte after the initial stages of growth. It is not, however, a persistent structure and represents the intermittent phase of the cycle. It is doubtful whether its omission even for a considerable period would seriously jeopardise the species. The sporophyte phase ultimately gives rise to spores which are formed by reduction division, so that on germination they restore the haploid phase. This sequence is referred to as the Alternation of Generations, and will be seen in a further degree in the Ferns.

It may be useful to conclude this section with a brief review of the differences between Liverworts and Mosses.

Liverworts	Mosses
Plant body may be thalloid or leafy.	Plant body always organised into stem and leaves.
Rhizoids unicellular.	Rhizoids multicellular.
Leaves in two rows, without midrib and underleaves present.	Leaves with a midrib, usually in three rows and without underleaves.
Capsule opening by four longitudinal valves.	Capsule opening by a lid and usually by peristome teeth.
Elaters present.	Elaters absent.
Protonema ill-defined and not always present.	Protonema typical.

Although individual items may be difficult to distinguish, the inspection of all the points enumerated should enable the specimen under consideration to be assigned to its group without difficulty.

3. PTERIDOPHYTA—THE FERNS AND THEIR ALLIES

In the Mosses and Liverworts we have seen the development of a simple plant body with stems and leaves, but without major internal differentiation.

In the Pteridophyta a further development is found and the plants have highly organised stems, roots and leaves with a well-developed vascular system. It must be emphasised that such structures cannot be regarded merely as a simple advance on the Moss condition because the typical Fern plant belongs to a different generation, the sporophyte. So it is much more difficult to trace the ancestry of the Ferns because it must be sought through the sporophyte, and the intervening links of fossil remains are not at present complete.

The Pteridophyta include the true Ferns or Filicales, the Clubmosses or Lycopodiales and the Horsetails or Equisetales. The general organisation of the Filicales will be discussed in some detail, but only a brief review will be made of the other groups, which today are represented only by a comparatively small number of herbaceous species, but which during the Palaeozoic period had almost a dominant place in the Plant Kingdom and produced many large and woody forms. In all the groups the life-history shows a complete separation into independent sporophyte which is the dominant, persistent phase and independent gametophyte which is usually insignificant, but which may also be rather long-lived.

The Ferns—Filicales

All the Ferns with which we are familiar or likely to be familiar can be described as herbaceous, and some of them are of very delicate structure. Nevertheless in tropical and subtropical regions a few Tree Ferns exist with well-marked woody stems.

The majority of Ferns are found growing in damp places, or at any rate in conditions in which a fairly high degree of humidity is maintained, though one common fern, Bracken, seems to be fairly tolerant of considerable exposure. In general they do not show much modification for resistance to desiccation, but of course the well-developed vascular system does mean that water can be supplied quite quickly to the leaves.

The typical Fern plant has a stout rhizome from which the leaves or fronds arise. In many cases these develop as a whorl immediately behind the apex, and usually there are two whorls developing at once—those of the current year and the rudimentary leaves of the following year. In the case of Bracken each rhizome branch produces only one leaf per year, although it is true that this leaf is a very complex one. Associated with the leaf base are numerous true adventitious roots which function in the same way as those of Flowering Plants. As the leaves die down the old bases are left behind, and in the common Male Fern, *Dryopteris filix-mas*, a stubby rhizome of considerable size is formed and much of it becomes exposed above the ground. In other Ferns like Bracken and Adder's Tongue the rhizome is completely subterranean and more slender.

The leaf or frond, which in most ferns is the only aerial structure, varies considerably in form. In Hart's Tongue Fern and in Adder's Tongue the leaf-blade is completely undivided. In *Polypodium* it shows a simple pinnate structure, whilst in *Dryopteris* (Fig. 68) this is further subdivided. Bracken has a single large leaf which is branched into a number of leaf-like segments. As a general rule the main divisions of the lamina are called **pinnae** and the subdivisions **pinnules**, whilst the main stalk is called the **rachis**. One of the most characteristic features of the fern leaf is the way in which the young leaves are coiled at the tip, a condition which is called **circinnate**. Frequently the young rachis bears many brown scales or **ramentae** which tend to fall off as the frond gets older.

In many Ferns only one kind of frond is seen, but in others a

special structure is formed to bear the spores, and reference
will be made to these later.

The internal structure of the various organs is much more
complex than anything seen in the previous groups. The
Ferns have a vascular system which in many respects is similar
to that of the Flowering Plants, although it differs in certain
important items.

A cross section of the rhizome of a Fern like Dryopteris
shows a more irregular arrange-
ment than in most Flowering
Plants, possibly because of the
closeness of the leaf-bases to one
another. The cortical tissue is
mainly parenchymatous, but
often there are bands of sclerenchymatous fibres. The vascular
tissues lie in the general parenchyma, and their organisation
varies in different Ferns. The
most primitive arrangement consists of a core of xylem surrounded by phloem and enclosed
by an endodermis. This is a
protostele. A further development shows a ring of xylem with
phloem both inside the ring and
surrounding it, and this is called
a **siphonostele.** Finally this
system may be broken up into
a series of " bundles ", each consisting of a mass of xylem sur-

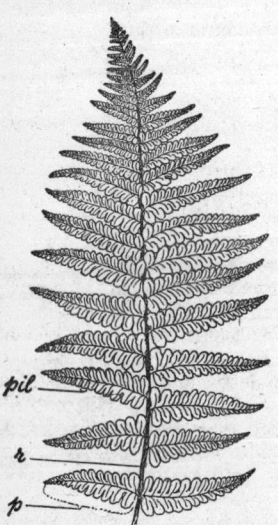

FIG. 68.—Frond of *Dryopteris*
(Male Fern).

p. pinna, *pil.* pinnule, *r.*
rachis.

rounded by phloem and enclosed
by the endodermis. This is the
condition seen in Dryopteris,
and is called a **dictyostele,**
each separate portion being a

meristele. The endodermis is very heavily thickened, and
in many cases consists of a double layer of cells. Within
the endodermis is a parenchymatous pericycle enclosing the
phloem. The central mass of xylem is mainly metaxylem with
irregularly placed groups of protoxylem. There is no cambium
at any stage.

The tissues themselves have certain differences from those
which were seen in the Flowering Plant. The phloem has no
companion cells and the sieve tubes are comparatively narrow,

with the sieve areas (which have very fine perforations) on the lateral walls. The pointed nature of the sieve-tube elements means that they have no true end walls. The xylem has no vessels, and the conducting elements are tracheids which show a typical scalariform thickening on the walls. These scalariform tracheids are typical of the Fern xylem.

A transverse section of the root shows a central stele which is typically diarch, a condition easily seen in Dryopteris. The endodermis is very strongly developed and the protoxylem is exarch, the arrangement more closely resembling that of a diarch Angiosperm root. Much starch is present in the cortical cells.

The general organisation of the leaf is rather similar to that of a shade leaf of an Angiosperm. A vertical section of the leaf shows a dorsiventral structure with stomata in the lower epidermis and palisade mesophyll cells below the upper epidermis (Fig. 70). An interesting feature is the presence of chloroplasts in the epidermal cells. Most Fern leaves are thin and have poorly developed cuticle. The vascular bundles are arranged as in most Angiosperm leaves, with the xylem towards the upper surface. In the rachis, however, the distribution of the meristeles is much like that of the rhizome. One point which must be made is that at the apices of the stem and root the meristem arises from a single apical cell which is tetrahedral in form. The various derivative tissues are produced by division from the sides of this apical cell.

In general, the Ferns show little vegetative propagation. Branches may arise from the rhizome and become separated, but there is no organised method, unless one is prepared to include under this heading the case of such plants as *Asplenium bulbiferum*, where new sporophytes grow out directly from the surface of the leaves. It must be pointed out, however, that this is really a special condition in which these small plants replace the sori.

The usual method of reproduction is that involving a spore-forming phase followed by a plant from which the gametes are developed. The spores are borne in sporangia, which may be found during the summer on the backs (dorsal surfaces) of typical leaves in many common forms. In *Dryopteris*, the Male Fern, many of the ordinary foliage leaves bear sporangia, and are therefore sporophylls, and this is also true of Bracken, Hart's Tongue Fern and many others. In the Hard Fern, *Blechnum spicant*, the sporophylls are more slender than the foliage leaves, whilst in the Royal Fern, *Osmunda*, the sporophyll in no way resembles the foliage leaves. In Adder's

Tongue there is a single spear-shaped foliage frond from the base of which there may or may not arise a slender sporophyll, the " tongue ".

Usually the sporangia are developed in clusters called sori, and in most cases the sorus is protected by a flap of tissue developed from the under-surface of the leaf—the indusium— the whole structure usually occurring over a vein. The indusium has different shapes in different groups; in Dryopteris it is kidney shaped (Fig. 69), in Hart's Tongue it is linear, whilst in Bracken the sori lie between the edge of the leaf and a linear indusium. The sporangia arise from a cushion of tissue called the placenta and are small box-like structures each with a stalk. In some Ferns they all mature together

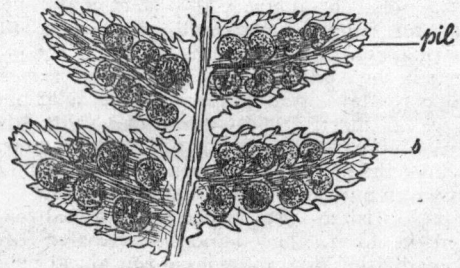

Fig. 69.—Pinnules of Male Fern with Sori.
pil. pinnule, *s.* sorus.

(*Dryopteris*), whilst in others a succession of sporangia can be seen in the same sorus.

Each sporangium arises as a small group of cells. These gradually differentiate to produce a wall and a central mass of cells from which arise the archesporium and tapetum. From the archesporial tissue the spore mother-cells separate, and they divide by meiosis to give the actual spores, which are unicellular and have a protective wall. In *Dryopteris* the stalk of the sporangium bears a small gland. The mature sporangium is a biconvex capsule in which the spores lie freely. The wall of the sporangium is very distinctive, having a band of thick-walled cells round the greater part of the edge. This is called the **annulus,** and at one point it is interrupted by a group of large, thin-walled cells which form the **stomium.** As the sporangium dries the tension in the walls of the annulus' eventually causes the rupture of the stomium and the annulus flies back, expelling the spores. These may be distributed by the wind. This principle of dehiscence is a general one in the

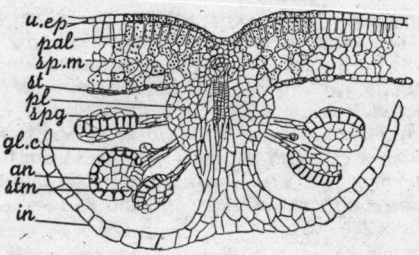

FIG. 70.—Vertical Section through Sorus of Male Fern.

an. annulus, *gl.c.* gland cell, *pal.* palisade mesophyll, *pl.* placenta, *sp.m.* spongy mesophyll, *spg.* sporangium, *stm.* stomium, *st.* stoma, *u.ep.* upper epidermis, *in.* indusium.

higher plants. A group of *Dryopteris* sporangia can be seen in Fig. 70.

The spores germinate on the ground, but the plant produced requires much damper situations than the sporophyte. This new generation is called the prothallus and varies somewhat in form, but in *Dryopteris* it is a thin heart-shaped organism varying in size from three or four millimetres across to about one centimetre (Fig. 71). The apical cell lies in the V of the prothallus, and the central region is thickened into a cushion of cells, from the under-surface of which, especially at the posterior end, a mass of rhizoids appears. From the cushion the prothallus becomes thinner to-wards the edges, where it is only one cell thick. The cells contain chloroplasts, and the structure is quite independent and may persist for some time. From the under-surface the sexual organs are produced. The male organs are antheridia and are found amongst the rhizoids and on the wings of the thallus.

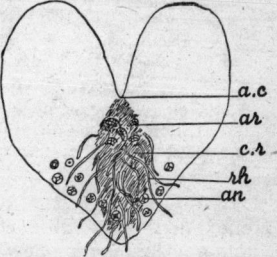

FIG. 71.—Diagrammatic View of Ventral Surface of Pro-thallus of Male Fern.

a.c. position of apical cell, *an.* antheridium, *ar.* arche-gonium, *c.r.* cushion region, *rh.* rhizoid

Each grows from a single prothallial cell and differentiates wall cells and antheridial cells. Each of the latter produces a single antherozoid or spermatozoid, which is coiled and bears a tuft of cilia at the narrow end. The antheridium is very small and produces only about a score of spermatozoids. Fig. 72 shows a section of the prothallus. The archegonia are produced towards the apex of the prothallus. Each has a venter sunk into the prothallial tissue and a short neck consisting of four rows of cells and curved towards the antheridia. There is an oosphere, a ventral canal cell and an indistinct mass constituting the neck-canal cells. At first the neck is closed, but when the archegonium is ripe the neck opens and exudes a liquid said

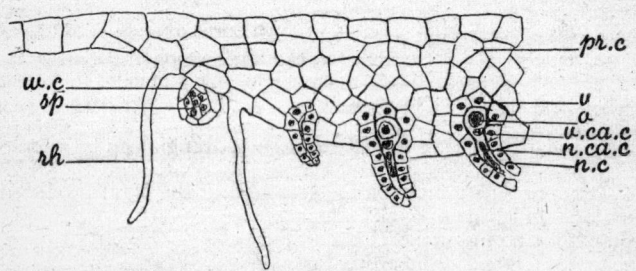

FIG. 72.—Vertical Section of Prothallus with Reproductive Organs.

n.c. neck cell, *n.ca.c.* neck canal cell, *o.* oosphere, *pr.c.* parenchyma cell with chloroplasts, *rh.* rhizoid, *sp.* spermatozoids, *v.* venter, *v.ca.c.* ventral canal cell, *w.c.* wall cell.

to be produced by the disintegration of the canal cells. In some Ferns this liquid contains malic acid, which causes a chemotactic movement of the spermatozoids which can be induced to enter a capillary tube containing malic acid.

It is important to realise that water is still essential for the movement of the spermatozoids, so that even though the Ferns have become well-developed vascular plants in one phase, they are very dependent on water in the sexual phase.

One spermatozoid fertilises the oosphere, and division into an embryo starts immediately. The oospore divides into two, then into four and then into eight cells, which are called **octants**. From these the future apices can be recognised, one producing the stem, one the leaf rudiment, a third the primary root, whilst the fourth apical cell gives rise to the foot, which grows into the tissue of the prothallus and absorbs nutrition for the young sporophyte. Gradually the young plant enlarges and the first leaf grows up near the apex of the prothallus. This leaf is usually very simple, and is the first of a succession which gradually develop the adult form. The young plant soon becomes independent with the formation of adventitious roots and the development of a small rhizome. It takes several years to reach the spore-bearing stage.

Thus the Ferns again show an alternation of generations, but in this case the sporophyte is the dominant plant and is long-lived. The continuity of the species is not affected by the failure of a crop of spores for one or even several years. The sporophyte is a long-lived vascular plant, whilst the gametophyte is merely a platform to bear the sexual organs, though in some cases it may last two or three years. The gametophyte is again the haploid generation and the gametes are produced by ordinary mitosis. Fertilisation of the oosphere restores the diploid condition. Most Ferns follow the above cycle, and there is a good deal of similarity in the form of the reproductive organs.

The remaining members of the Pteridophyta are represented by the comparatively small plants seen in the Horsetails (Equisetales), the Clubmosses (Lycopodiales) and the Selaginales.

The Horsetails are peculiar-looking plants in which the leaves are merely whorls of scales on a simple or branched aerial stem which grows up from a rhizome. The latter also has scale leaves and adventitious roots. The stem is green and the shoots are annual. They have a well-developed vascular system and the internal structure has many of the features of aquatic plants, though many modern Horsetails are not aquatic. Reproduction follows the same pattern as in the Filicales. The sporangia, however, are borne in heads or strobili which may terminate the ordinary shoots or appear on separate shoots. Each has a number of small sporophylls, under which are the sporangia. The spores when released are all alike, but some produce male prothalli and some female prothalli. Fertilisation again depends on water. The Horsetails are common in this country.

The Lycopods are somewhat Moss-like plants often with trailing stems and small pointed leaves. The vascular system is central and rather simple. Some of the shoots are erect, and they bear strobili in which the sporangia arise on the upper side of sporophylls resembling ordinary leaves. The sporangia are all alike, and the spores germinate to produce a peculiar fleshy prothallus which is saprophytic, having no chlorophyll, and bearing the organs on the upper surface. The British Lycopods are mainly upland plants.

There is one British species of Selaginella, *S. spinosa*, which is found in upland regions, though a number of trailing species are cultivated. The leaves are very small and the shoot develops peculiar rooting structures called rhizophores. The whole plant is again moss-like.

The sporangia are borne in erect strobili and are of two kinds, so that, in contrast to the **homosporous** condition we have so far encountered, we now see **heterospory**. The micro-sporangia usually occur at the upper end of the strobilus, and the small spores when released germinate to give a very simple prothallus which is mainly an antheridium producing biciliate spermatozoids. The megasporangia are borne on sporophylls in the lower part of the strobilus, and each produces only four megaspores. These start to germinate before leaving the plant and later develop into very reduced prothalli on which are the simple archegonia. Neither male nor female prothallus has any chlorophyll, and they live only long enough to produce the sexual organs. Fertilisation takes place as in Ferns and a new sporophyte appears.

It has already been mentioned that these plants are very reduced forms of the large tree-like types which existed in earlier periods. In most cases the internal structure suggests the aquatic origin of these groups.

4. GYMNOSPERMS

Again we find a gap in the line of development from the Pteridophyta to the Spermatophyta, especially in the evidence of existing species. This gap is partly closed by the fossil record, but the origin of the Spermatophyta is far from clear.

The Gymnosperms are Spermatophyta in which certain differences are apparent compared with Flowering Plants. The most familiar Gymnosperms are the Conifers, which include Pines, Spruces, Larch, Cedars, Cypress, etc., and are character-ised by the possession of evergreen leaves which are either needle-like, as in Pines, Spruce, Larch, Cedar, etc., or small and closely packed, as in Cypress and *Sequoia*. All the Conifers are trees, and some of them attain enormous size—the giant redwoods being more than 300 feet high. Conifers of various kinds are found in all parts of the world, and par-ticularly in colder latitudes or at higher altitudes, as they are very hardy. Many of them provide valuable timber, and in this country we are familiar with extensive Conifer plantations in the State Forests.

The Scots Pine, *Pinus sylvestris*, is a native of Britain, and its structure and life-history give a fairly general picture of the Conifer organisation.

If a young tree is examined it will be seen that the branches arise in whorls and obviously start from a group of buds which can be seen near the apex of each first-year branch. As the tree grows older this arrangement is not so obvious and often the lower branches are lost, leaving a bare trunk with a crown of branches at the top. Most of the Conifers exhibit the regular branching habit, and it will be noticed that with few exceptions there is a very dominant main trunk, so that most Conifers are very symmetrical in appearance. The Yew and the Juniper are native trees which do not show this habit.

In the younger branches the bark tends to be fairly even, though not smooth, but as the tree gets older the bark becomes hard and patchy. The leaves are very distinctive. They arise in pairs in Scots pine, and each pair actually represents a short shoot or dwarf shoot, as it is called, and small-scale leaves envelop the base of the true leaves (Fig. 73). The leaves are long, tough and needle-like

FIG. 73.—A. Shoot of *Pinus* with Male Cones. B. Single Male Cone.

d.sh. dwarf shoot, *m.c.* male cone, *n.l.* needle leaves, *n.sh.* new shoot, *m.st.* main stem, *sp.* sporophylls.

with a semicircular cross-section (Fig. 76). In other species of *Pinus* the number of leaves on each dwarf shoot may differ, but it should be noted that in the young seedling of all the species the first true leaves arise directly and individually from the main axis.

In Spruce the foliage leaves arise directly all round the twig, in Cedar and Larch they arise in clusters from short lateral shoots, whilst in Fir and Yew the leaves are more flattened and tend to be arranged in two rows (though they are actually inserted all round the stem). In most cases the leaves stay on the tree for two or three years or rather longer, whilst in *Araucaria*, the Monkey Puzzle, leaves have been found which were forty-five years old and still living. On the other hand, in Larch the leaves are shed each year. The leaves of most Conifers are dark green and give a familiar appearance to the woods.

The seedling Pine has a number of cotyledons (up to twenty-four) which are epigeal and act as foliage leaves until the true

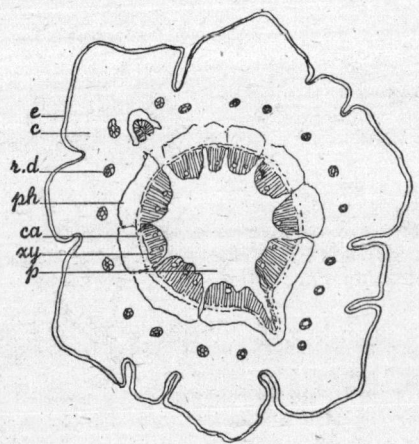

FIG. 74.—Diagrammatic Cross-section of One-year-old Stem of
Pinus.

c. cortex, *ca.* cambium, *e.* epidermis, *p.* pith, *ph.* phloem, *r.d.* resin duct, *xy.* xylem.

leaves appear, first singly and soon as the typical short-shoot arrangement. The roots form a fibrous system, and at an early stage many of them form an association with a fungus, giving an external or ectotrophic mycorrhizal system. Such roots are short and stubby and are easily recognised.

The general anatomy of the stem as seen in transverse section is not unlike that of a dicotyledon twig. In the seedling there is a ring of vascular bundles with internal protoxylem, external metaxylem, cambium and phloem. At an early stage a complete ring of wood is produced, and a section of a one-year-old twig has the general appearance shown in Fig. 74. A thickened epidermis encloses a parenchymatous cortex in which there is a ring of resin ducts. Each resin duct is a canal enclosed by a wall of secretory cells and supporting cells. The secretory cells produce the aromatic, sticky resin which is such a feature of these plants. Resin ducts are also present in the vascular tissues.

The vascular tissues are distinctive. The xylem consists of elongated narrow tracheids only—there are no vessels in Conifers. In section the tracheids are seen to be very regular

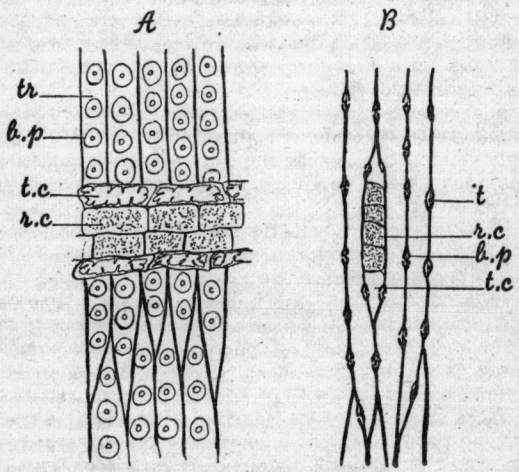

FIG. 75.—A. Radial Longitudinal Section of Wood of *Pinus*.
B. Tangential Longitudinal Section of Wood of *Pinus*.

b.p. bordered pit, *r.c.* ray cell (parenchymatous), *t.c.* tracheidal cell of ray, *tr.* tracheid, *t.* torus.

in arrangement and vary little in dimension between spring and summer wood. On the radial walls there is a row of bordered pits which are characteristic of Conifers (Fig. 75A). In tangential and transverse sections of the wood the pits may be seen in profile (Fig. 75B). Since there are no vessels and there is no direct communication between the tracheids, the contents of the latter must presumably be transported by passage through the thin portion of the pit wall. The central part of this region is thickened (the torus), and if the pit wall is pushed over to one side it can close the pit opening in the secondary wall.

The wood is penetrated by very narrow medullary rays rarely more than one cell wide. They are of special interest because the marginal cells (i.e. above and below the actual ray cells) are tracheidal and have bordered pits in the xylem region, where they probably take part in lateral translocation.

The cambial cells are very narrow and elongated, so that neither they nor their products have any true end walls.

The phloem possesses only sieve tubes and parenchyma, companion cells being absent. The sieve tubes are very narrow and the sieve areas are on the radial walls. In the phloem the tracheidal cells of the medullary rays are replaced by albuminous cells which may be storage tissue or serve for lateral translocation. Starch grains are frequent in the rays, particularly in the xylem.

Secondary thickening is carried on as in the Dicotyledons, and annual rings are easily recognised in the wood, though it is probably generally true that the rate of growth, and therefore the amount of wood produced in the season, is less than in Dicotyledon trees.

A cork cambium arises in the cortex during the first year's growth, and cork is produced, cutting off the epidermis from the cortical cells below.

The anatomy of the leaf is very striking. The central region is occupied by a stele in which two small vascular strands are embedded in a mass of parenchyma with some fibres appearing as the leaf gets older. The stele is bounded by a well-defined endodermis with prominent Casparian strips. The xylem faces the flat (ventral) surface of the leaf, and outside each xylem group is a patch of tracheidal cells which act as transporting tissue. Albuminous cells are associated with the phloem. The amount of vascular tissue is very small for the size of the leaf and there are *no* veinlets. Although the leaves may remain on the tree for three years, it seems that there is little further production of vascular tissue except for some phloem in the second year. A great deal of starch is stored in the leaves. Fig. 76 shows a vertical section of the leaf.

The mesophyll is a mass of parenchyma cells with peculiarly infolded walls, a condition which presumably increases the surface area for diffusion without making excessive air spaces to increase transpiration. All the mesophyll cells contain chlorophyll, but there is no differentiation into palisade and spongy tissue. Resin ducts are present. The leaf is enclosed by an epidermis with sclerised walls and supported by a hypodermis similarly strengthened in which the stomata are situated. As in the Flowering Plant, the guard cells contain chlorophyll, but their presence in the hypodermis means that the stomata are placed at the bottom of a small epidermal pit, an arrangement which will cut down transpiration and adds another xerophytic character to the leaf.

The root is usually diarch but may be pentarch and is very similar to that of a Dicotyledon in arrangement of the tissues. The xylem groups are Y-shaped with the metaxylem internal and a resin duct between the protoxylem arms of the Y. A cambium arises as in the Dicotyledon root and secondary

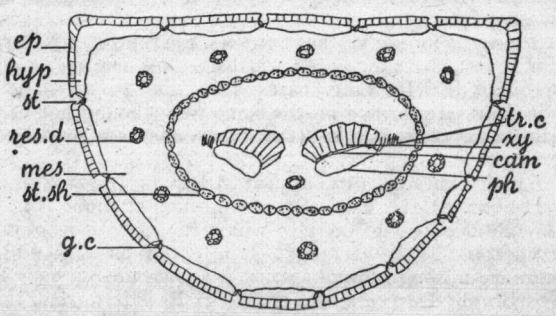

FIG. 76.—Diagrammatic Cross-section of Leaf of *Pinus*.

ep. epidermis, *cam*. cambium, *g.c.* guard cell, *hyp*. hypodermis, *mes*. mesophyll, *ph*. phloem, *res.d.* resin duct, *st*. stoma, *st.sh.* starch sheath, *tr.c.* tracheidal cell, *xy*. xy.em.

thickening proceeds in the usual way. A special feature is the development of the mycorrhizal roots, and these organs have few or no root-hairs.

It should be noted that at both stem and root-apices of the Conifers there is an apical meristem and not an apical cell, as in the lower plants.

These anatomical features are fairly representative of general Conifer structure, though individual differences do occur. Thus in *Taxus* and others there is a tertiary spiral of thickening on the tracheids, and again in Taxus the leaf is less xerophytic and also rather more dorsiventral in appearance and structure.

Reproduction

The Conifers show very little natural vegetative reproduction and indeed are very difficult to propagate vegetatively in cultivation.

On the other hand, there is a very definite and highly

G

developed sexual reproduction which results in the production of a new structure called a seed. The seed contains an embryo, a store of food material for the developing seedling and a protective covering which enables the seed to withstand adverse conditions and makes it possible for germination to be delayed until circumstances are suitable to support growth. This is a condition which is not found in any of the lower groups, where development of the zygote must usually follow fairly quickly on its formation and it is the development of the seed habit which has undoubtedly led to the dominance of the Spermatophyta.

In *Pinus*, as in many other Conifers, the reproductive organs are borne in cones which are specialised axes bearing a number of sporophylls. In many cases, including *Pinus*, male and female cones are formed on the same tree, though not usually on the same twig, but in *Taxus* the cones are produced on different trees.

The male cones of *Pinus* appear in May or June and replace dwarf shoots at the base of a new principal shoot (Fig. 73). Each cone is an ovoid body in which the male sporophylls or microsporophylls are arranged spirally on an axis and the appearance in longitudinal section is shown in Fig. 77. Each microsporophyll is a small flat scale with its external end turned upwards. The microsporangia or pollen sacs are not visible externally and are arranged longitudinally side by side on the underside of the microsporophyll. Two pollen sacs are present on each scale. Each has a wall several cells thick, and within this pollen mother cells are formed which undergo reduction division, and the resulting cells become the microspores or pollen grains. Each has a double wall, and in Pinus the outer wall or exine is blown out into two balloon-like extensions which become filled with air and are used as floats in the distribution of the pollen grains. The pollen grain starts to germinate before it is shed. The original nucleus divides and a small group of cells is produced against the wall of the grain. The lowest cells are short-lived, but the terminal cell forms the antheridial cell. The remaining nucleus and cytoplasm form the tube cell and subsequently undergo further growth. At this stage the grains are shed by the splitting of the wall of the pollen sacs due to uneven contraction caused by the drying of a fibrous layer—a mechanism largely similar to that operating in the fern sporangium and also in the pollen sac of the Flowering Plant. The pollen grains are wind-distributed.

The female cones (Fig. 78) are produced at the apex of a new main shoot, and each replaces one of the buds which normally produce a new lateral shoot. In *Pinus* the cones are small, tightly closed structures and are purplish or dark red in colour. The megasporophylls are spirally arranged on the axis of the cone and each is formed from a pair of scales, the lower scale being smaller and known as the bract scale, whilst the upper scale is called the ovuliferous scale. It is on the exposed surface of this scale that the two ovules are borne—this being

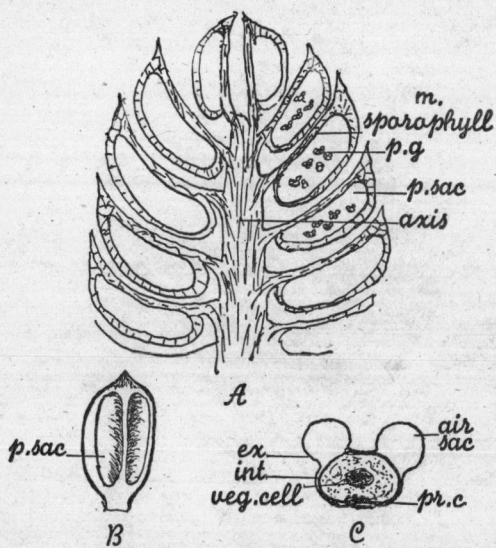

FIG. 77.—A. Vertical Section of Male Cone of *Pinus*. B. Single Sporophyll. c. Pollen Grain.

ex. exine, *int.* intine, *m. sporophyll.* microsporophyll, *p.sac.* pollen sac, *p.g.* pollen grain, *pr.c.* prothallial cells, *veg. cell* vegetative cell.

the origin of the name " Gymnosperm ". The ovuliferous scale is much larger than the bract scale and increases in size during the development of the cone. The arrangement is not constant in all conifers, but many of them have this organisation.

Each ovule is formed at the inner end of the ovuliferous scale, and has a single wall or integument with an opening or micropyle facing towards the axis. Within the wall there is a region called the nucellus with its apex at the micropylar end and free from the integument. In this end of the nucellus a megaspore mother cell is formed, and by reduction division produces four megaspores, but only one of these develops to form the true megaspore. This enlarges rapidly, presumably at the expense of the nucellar tissue, and it produces what is in fact an internal prothallus, a gametophyte wholly contained within the sporophyte parent. It is provided with a good deal of food material, and at the micropylar end two or three

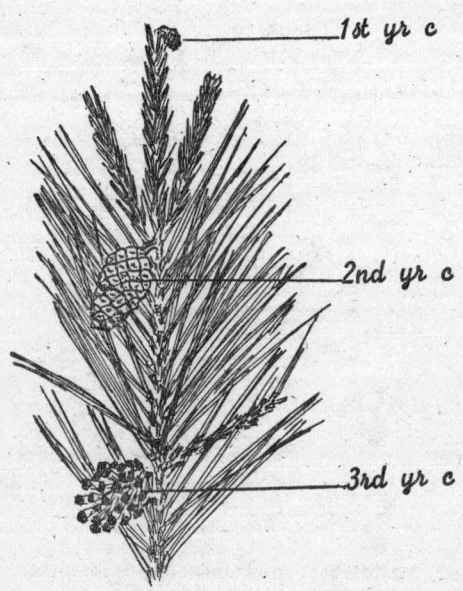

FIG. 78.—*Pinus* Shoot with Female Cones.

1st yr. c. 1st-year cone, *2nd. yr. c.* 2nd-year cone, *3rd. yr. c.* 3rd-year cone.

archegonia appear, each consisting mainly of a venter with a large oosphere, beyond which is a short neck with a single degenerate canal cell. An ovule is shown in section in Fig. 79.

The pollen, which is produced in large quantities, is blown by the wind on to the female cones, the scales of which now open, so that the pollen becomes lodged inside the cone near the micropyle of the ovule. After pollination these scales close again. The pollen grain then becomes enclosed in a drop of fluid secreted from the nucellus, and as this fluid dries the drop shrinks into the micropyle carrying the grain into contact with the apex of the nucellus. Germination of the grain now continues for some time, the antheridial cell producing two male nuclei, whilst the tube cell pushes out the inner wall of the pollen grain to form the pollen tube. This penetrates a short distance into the nucellus and then stops, and in *Pinus*

FIG. 79.—Longitudinal Section of ovule of *Pinus*.

ar. archegonium, *br.sc.* bract scale, *int.* integument, *mic.* micropyle, *nuc.* nucellus, *pr.* prothallus, *ov.sc.* ovuliferous scale.

no further progress towards fertilisation occurs in that season, although the female cone continues to increase in size and becomes woody. (In some Conifers, e.g. Larch, the whole process is completed the same year and the seed is ripe in the autumn.) In the following spring the pollen tube recommences its growth through the nucellus towards the prothallus. On reaching the prothallus the tip of the pollen tube bursts, releasing the two male gametes, which are not spermatozoids, and these enter the archegonium. Only one is concerned in fertilisation—a marked difference from the condition in the Flowering Plant—but it is said that in some Conifers the second male gamete may fertilise another oosphere.

Division of the zygote starts immediately and two successive nuclear divisions produce four nuclei, which pass to that end of the zygote away from the micropyle. A number of divisions then follow, so that four columns of cells are produced, and in *Pinus* these may separate at the inner end so that there are in effect four suspensors, each with an embryonic cell at the free (prothallial) end. The embryonic cell continues to divide as it is pushed into the prothallial mass and eventually becomes lobed as the numerous cotyledons begin to differentiate, together with the radicle and plumule.

The prothallus is the endosperm, and as the embryo develops it absorbs some of the material, but when the seed is ripe there is still sufficient endosperm to enclose the embryo. The endosperm contains a good deal of oil. Although several proembryos form, only one normally matures, a condition which is true of most Conifers.

In the meantime the integument has hardened and become the seed coat or testa, and from the surface of the ovuliferous scale a wing is formed which is attached to the testa. The cone continues to increase in size, and growth goes on until the late summer of the third year, when the scales open, probably by drying, and the seeds are released to be blown about by the wind. Thus the whole cycle takes nearly two and a half years.

Although the cone structure is characteristic of Conifers it is not the only reproductive arrangement. In *Taxus* (Yew) the ovule is single in a small bud and the seed becomes enclosed in a bright pink fleshy aril, whilst in Juniper the seeds are embedded in a black or dark blue berry-like structure. In both cases the seeds will be distributed by birds which eat the fleshy part.

The most important part of this reproduction is that water plays no part in the distribution of the gametes.

The other Gymnosperms are of much less wide distribution than the Conifers.

The Cycads are somewhat palm-like plants of tropical and subtropical regions. The group includes such genera as *Cycas*, *Zamia*, *Encephalartos* and others, and most have large feathery leaves arising in a crown from the top of the stem. In some cases the young leaves show a fern-like coiling. The plants produce rather typical male cones in which the microsporophylls have numerous small sporangia on the under side. On the other hand, the megasporophylls are rather like small foliage leaves with the ovules attached laterally at the base. The microspores or pollen grains are blown by the wind and are drawn into a pollen chamber at the apex of the ovules in the same way as the pollen grain is drawn into the Pinus ovule. In the Cycads, however, the microspore on germination produces two large top-shaped spermatozoids with spirally arranged cilia, and one of these effects fertilisation. The ovule produces a large seed not unlike a horse chestnut or hard plum.

Another Gymnosperm, the Ginkgo or Maidenhair tree, also produces spermatozoids.

Finally there are a few shrubby Gymnosperms forming the Gnetales and including one of the most peculiar plants in the world—*Welwitschia mirabilis*—like a large beetroot buried in the ground and producing only two strap-shaped leaves in its whole life and found only in a small area in South-west Africa.

These genera are probably the most advanced Gymnosperms and the wood possesses true vessels.

4

CLASSIFICATION

AN outline classification of the Plant Kingdom has already been referred to, but it is necessary now to say a little more about the arrangement.

The classification of organisms is not easy, and many attempts have been made to produce a satisfactory system. The idea of classification is to provide a means of reference to the individuals and also to crystallise the relationships between the various types. Obviously if we have an ordered system of known forms, the chances of correctly placing new individuals will be greatly increased.

Virtually all parts of the plant structure are used in classifying plants, but in the lower groups it is probably true to say that the vegetative characters are less distinctive individually, though probably sufficiently characteristic to identify the group immediately.

At the other end of the scale the Flowering Plants are divided into families on characters which are mainly associated with the reproductive organs, whilst closer identification brings in vegetative differences to supplement the floral characteristics. Such identification is not always easy and indeed in some families it is extremely difficult, so that the isolation of some species is very much the job of an expert. In order to enable identifications to be carried out, various systems of tabulated data have been produced in the form of keys, by which, starting from common characters, the botanist is able to eliminate family after family, genus after genus, species after species, until at last he is left with the individual under consideration. Even then there may be many varieties within the species. This pyramid of relationship is based on the **species**—regarded as the lowest whole unit in the system. It is not easy to define what is meant by a species, though a general description is that it is a unit amongst the members of which there is free interbreeding. In plants, however, it is not certain that this would impose the limits which other considerations require. Species differ in minor characters—frequently vegetative, but persistent and regular, and it must be confessed that in plants much confusion can arise in some families.

Species having certain characters in common are grouped

into a **genus,** and genera are separated by more marked differences, mainly floral, or at any rate reproductive. The genera in turn are grouped together to form a **family** (if this is very large there may be sub-families), in which certain major features, such as floral insertion, numbers of stamens, type of ovary, etc., are characteristic throughout, though a deviation in any one feature may be outweighed by the general agreement of other critical points. Above this level certain families may be included in a **cohort.**

An individual plant normally has two Latin names, the first being the generic name and the second the specific name. Thus the Primrose is *Primula vulgaris*, whilst the Cowslip is *Primula veris* and the Oxslip is *Primula elatior*. The specific name begins with a small letter, and the name of the plant may be followed by the letter L, denoting Linnaeus, or some other abbreviation if the name is attributed to another botanist. It is customary to adopt the first known name of the plant for general use. It may be added that nomenclature is often a vexed question, and in the lower orders particularly names change quite frequently. It is not unusual to find that a name which has been accepted for many years has to be changed because an earlier, different one has been unearthed. The same general rules of nomenclature apply to all plant groups, but identification and agreement on names is even more difficult in the lower forms than in Flowering Plants.

It is not proposed in this book to give a detailed summary of the various families of Flowering Plants. Reference is made to suitable works which give all the necessary information.

The description of the flower is often aided by special signs and abbreviations which enable the information required to be expressed in one line, whilst a plan or floral diagram and an elevation or vertical section show the general relationships of the floral parts.

Fig. 41A shows the floral diagram and vertical section of the Meadow Buttercup, whilst underneath is the floral formula. This is repeated below with its interpretation.

$\oplus$, $\male\female$, $\overline{\text{K}}5$, C5, A∞, $\underline{\text{G}}\infty$.

$\oplus$ Flower regular (actinomorphic).

$\male\female$ Flower hermaphrodite.

$\overline{\text{K}}5$ Five free (polysepalous) inferior sepals.

C5 Five free (polypetalous) petals.

A∞ Numerous free (polyandrous) stamens.

$\underline{\text{G}}\infty$ Numerous free (apocarpous) carpels which are superior.

By contrast we may consider the description of the flower of the White Deadnettle shown in Fig. 41B.

$\dagger$, $\male female$, $\overline{K}(5)$, C(5) A2 + 2, $\underline{G}(2)$.

This is the formal formula and its interpretation is:

$\dagger$ Flower irregular (zygomorphic).
$\male female$ Flower hermaphrodite.
$\overline{K}(5)$ Five sepals, inferior, joined to form a tube (gamosepalous).
C(5) Five petals joined together (gamopetalous).
A2 + 2 Four stamens in two groups, attached to the petals (epipetalous).
$\underline{G}(2)$ Two carpels joined together (syncarpous) and superior.
Details of placentation are shown on the floral diagram.

These are just two illustrations of the many types which may be encountered, but they give an idea of the various abbreviations.

A brief classification of all plant groups follows.

THALLOPHYTA:

   (i) Algae.

       (a) Chlorophyceae.   Green Algae.
       (b) Phaeophyceae.   Brown Algae.
       (c) Rhodophyceae.   Red Algae.
       (d) Cyanophyceae.   Blue-green Algae.

  (ii) Fungi.

       (a) Phycomycetes $\begin{cases} \text{Oomycetes.} \\ \text{Zygomecetes.} \end{cases}$
       (b) Ascomycetes.
       (c) Basidiomycetes.
       (d) Myxomycetes.   Slime Fungi.
       (e) Schizomycetes.   Bacteria.

 (iii) Lichens.

BRYOPHYTA:

   (i) Hepaticae.   Liverworts.

       (a) Marchantiales.
       (b) Jungermanniales.
       (c) Anthocerales.

(ii) Musci.   Mosses.

    (*a*) Sphagnales.
    (*b*) Andreales.
    (*c*) Bryales.

PTERIDOPHYTA:

  (i) Filicales.   Ferns.
 (ii) Equisetales.   Horsetails.
(iii) Lycopodiales.   Clubmosses.
 (iv) Selaginales.   Selaginellas.
  (v) Isoetales.   Quillworts, etc.

SPERMATOPHYTA:

I. Gymnosperms.

  (i) Cycadales.   Cycads.
 (ii) Ginkgoales.   Maidenhair Tree.
(iii) Coniferales.   Pines, Spruces, Cedars, etc.
 (iv) Gnetales.   Gnetum, Ephedra, etc.

II. Angiosperms.

 (i) Dicotyledones.

    (*a*) Archichlamydae (Polypetalae).
    (*b*) Metachlamydae (Sympetalae).

(ii) Monocotyledones.

5

PLANT COMMUNITIES

IN the preceding pages some attempt has been made to describe the structure and activities of individual plants or representatives of groups of plants. It will, however, be readily appreciated that very rarely, except under artificial and cultivated conditions, do we get a single species or variety growing in masses and without competition.

Under natural conditions various species grow together, and must therefore have some effect on one another. In addition, soil conditions play a great part in determining the type of plant in a particular area. The study of such conditions and the relationships involved is known as **ecology**.

Thus, as we look round at the vegetation of our own neighbourhood or of the British Isles and over even wider boundaries, it is found that there are various types of habitat which have a distinctive flora.

To a great many people the most familiar scene is the presence of large cultivated areas with their crops sown and developed by man. Even here, however, it is possible to see some natural association in the hedgerows which surround the fields and in the weeds which spring up with the growing crops.

It has already been suggested that the nature of the soil plays a considerable part in determining the type of plant which will grow on that soil. This is particularly the case in the early colonisation of an area. Thus when a field is ploughed and made ready for sowing, it is open to colonisation by any type of plant of which the seeds can reach the area or of which vegetative organs can enter from the margins. It is true that most of the species entering will be from neighbouring areas, and thus already suited to the environment, but it does not preclude the arrival of seeds from species growing further away and in different conditions.

When the seeds germinate, those species which can utilise the new ground will obviously grow more vigorously than those which cannot, and the latter will probably die off or be overgrown by the more successful species. Thus in time the ground will become covered with vegetation, and gradually a few species will dominate the colony, either because of rapid growth, or because they are perennials with reserves of food materials, or because they are large plants which can reach the light.

This is the kind of sequence seen in gardens or other cultivated land where the crop plants usually dominate by reason of numbers (assisted very often by weed removal by man) and the weeds are kept to the margins to compete with one another.

In such cases it is rare to find the area supporting its maximum number of inhabitants—it is still possible for others to find a place, and such a community is said to be open. If it were left, however, we should find that a succession of species would develop, and eventually perennial forms, probably grasses, would dominate the area, at least for a time, followed by tree seedlings, giving rise to scrub and eventually rough woodland. This would take a long time, and with the appearance of bushes and trees the amount of light at ground level would decrease, there would be greater competition for water and salts from the soil, so that we should find the number of herbaceous plants decreasing. Most people are familiar with the fact that fewer herbaceous plants are found in dense woodland than in open country.

So in time the plant succession would tend to become stabilised and the community closed. A definite relationship would have been established between the various types of plants, and levels of plant types could be recognised.

Such a development is called a **climax association**, and it can be seen in some very old woodlands, though normally it is prevented by various factors.

But competition amongst the species themselves and actual physical overcrowding are not the only factors which may lead to a closed community.

In many areas the soil is poor either due to lack of available water, as in sand dunes and shingle regions, or there may be a shortage of mineral salts due to leaching, or there may be failure of the bacterial cycles of breakdown because the soil is too acid. Such habitats are frequently very exposed, so that there is another physiological restriction to the growth of plants in the loss of water by evaporation. In such places the density of the plants will be determined by what the soil can support and the nature of the species by their adaptability to the conditions. Here we get a special illustration of what was mentioned earlier. Such habitats as sand dunes and acid moorlands are frequently within the range of wind-blown seeds of all kinds. Such seeds may germinate, but many of them cannot survive, and so the species are not represented in the habitats mentioned. Such communities are physiologically closed. (It may be noted that behind the sand dunes pockets frequently develop in which water can collect, so that some species can start to grow. These then bind the soil and form more organic material, and gradually such dune-slacks become rich plant communities, especially as they are frequently basic in soil reaction.)

The acidity or basicity of the soil is one of the primary conditions in determining the types of plant which may be found. In general, a base-rich or chalky soil supports a much wider and richer plant population than an acid one. This is not only due to the reactions of the plants themselves, but also to the fact that the activities of the micro-organisms are often restricted in the more acid or waterlogged soils, so that there is a poorer return of nutrients to the soil and the latter is therefore less able to support a heavy plant population.

In general, it is not difficult to diagnose the soil reaction by a brief glance at the dominant flora. Though some species are tolerant and will appear in a variety of environments, others are barred from some habitats by the soil conditions.

Thus an acid soil will frequently be indicated by the presence of such species as Heather, Bilberry, Birch, Moor Mat Grass, Wavy Hair Grass, Purple Molinia, many Sedges, the Hair Moss (*Polytrichum*) and Pines, and in fact some of these will always be represented. Groupings of these plants are characteristic of heaths and moorlands. Both have acid soils poor in nutrients, but in general heaths are drier, with sandy, gritty soils and rather shallow peat, whilst moorlands tend to be wetter and are usually underlain by deep peat. Some Flowering Plants, as exemplified by Woodsage, Heath Bedstraw and Eyebright, will appear on heaths more frequently than on moorlands, whilst in the wetter parts of the moorland the general species will be accompanied by Crowberry, Cranberry, *Sphagnum* moss and Sundews.

Many of these plants reflect their environmental conditions in their structural modifications, and most of them have morphological and anatomical devices to prevent water loss, whilst their roots often have mycorrhizal associations to overcome the difficulties of absorption.

In areas where the soil is basic there is usually a much wider range of species and the growth is more luxuriant. Basic soils are much richer in nutrient materials and, especially when chalky, are much warmer than acid soils, possibly because they are much better aerated.

Limestone hills and chalk downs always show a rich profusion of species, and on the whole are possibly the most interesting and rewarding areas for the botanist interested in Flowering Plants. Characteristic plants of basic soils are Rockrose, Hoary Plantain, Salad Burnet, Carline Thistle, Burnet Rose, Quaking Grass, *Brachypodium*, Autumn Gentian, Field Scabious, Ash, Spindle and Beech, to which must be added the species so frequently found on limestone, some of which will be mentioned in connection with Ashwood flora.

If we compare this list with that given for acid soils it is found that no species is common to both lists, though in an exhaustive census some plants would be found occurring in both areas.

Many species are widely distributed and are tolerant of all but the most extreme conditions. Thus it is possible to find Dandelions, Nettles, Daisies and grasses like Cocksfoot and Ryegrass in a wide variety of habitats, though their size and vigour may be very much affected.

The whole question of plant ecology is an intricate one, and the general situation is much confused by the activities of man. Thus there may be in one's own vicinity a typical Oakwood on more or less neutral soil. The dominant tree will be Oak, very often accompanied by Ash and frequently Wild Cherry. Such woods are fairly open, admitting a considerable amount of light, and there is often a shrub layer of Hazel with Honeysuckle and perhaps Dogwood, whilst the herb layer will include Bluebell, Bracken, Dog's Mercury, Ramsons, Creeping Soft Grass with Speedwells, Yellow Pimpernel and mosses on the ground surface. That represents a fairly stable condition. But as soon as the trees are felled the herb layer increases in quantity and luxuriance. Red Campion and Willow Herb, frequently Thistle and on open ground Wild Strawberry rapidly increase, often to be crowded out later by Blackberries and Bracken. The relatively open woodland becomes a dense scrub, but if Bracken does not take over, it is often found that in due course the Oakwood is succeeded by Birchwood and the neutral soil condition tends to become acid, with a decrease in the flora, even though there may be more light than in the Oakwood.

At this point it may be appropriate to mention the other types of woodland which are to be found. In a few areas, especially on chalk, Beechwoods occur. These are characterised by a close, leafy canopy through which little light is able to penetrate to ground level. This, together with the shallow rooting of the Beech, results in a poor ground flora, and the ground is often nearly bare of herbaceous plants, but certain parasitic or saprophytic plants can grow, and there is generally a rich fungus flora. This effect on the ground flora can usually be seen even when there is only a single Beech tree, provided it has a fairly large spread.

Old Ashwoods are often found on limestone. They are very open, in fact there can hardly be said to be a continuous tree layer, and there is a wide variety of shrubs and herbaceous plants amongst the Ash trees. In such situations Hazel, Bird Cherry and Juniper are found in the shrub layer with Lily of the Valley, Solomon's Seal, Ramsons, Bluebells, Campions, Wood Anemone, Dog's Mercury and less common species among the herbaceous types.

Conifer woods are common, but most of them are planted so that the trees are close together and little light penetrates, the ground often being bare except for Fungi. Older Pine forests are still represented in Britain, and these are more open, with Heather, Bilberry and various Ferns and Mosses in the herbaceous layer.

It must be pointed out, however, that much of the woodland in this country is mixed and the result of planting. In such woods one may find Ash, Oak, Beech, Sycamore and even Horse Chestnut.

So in most parts of the country it is difficult to find an original association—even on the moors burning of the Heather will produce some changes, though the constant soil conditions usually lead to the same plant associations in a short time. Nevertheless whatever the interference there still remains a plant relationship, and the study of the changes is of great interest.

The flora of the sea-coast has special points of interest. In the rocky areas regularly covered by the tide and below low tide level the flora is almost entirely algal, though in some flatter estuarine regions there may be found the Flowering Plant *Zostera*. Apart from this, the coastal zones are usually shingle beaches, sand dunes and salt marshes. The first two represent communities in which the soil is loose and through which water rapidly drains, whilst the plants are often exposed to salt spray. The species found there have deep root systems and many have fleshy leaves capable of storing water, whilst others have structural modifications for the reduction of water loss, as for example the rolling of the leaves in Marram Grass, Sea Lyme Grass and others or by the reduction of leaf surface. The species are usually scattered and the community is physiologically closed. Sea Holly, Sea Purslane, Sand Convolvulus, Sea Rocket, Sea Couch Grass and Marram Grass are all plants which may be found in these situations. In most cases these habitats merge inland to pasture, etc.

The salt marsh is rather different. Here there may be frequent inundation, and there is often a graded change from mud-flats to a proper land environment. The soil is usually saturated with salt, except after heavy rains and the plants live in a state of physiological drought. Fleshy leaved species are again common, and are exemplified by Glasswort, Sea Blite, Sea Lavender, Sea Aster, Thrift and Sea Arrow grass.

In all cases the landward side of the habitat will come to support species which can be regarded as typically inland forms, and common examples are Birdsfoot Trefoil and various Clovers.

Another important habitat of which little has yet been said is the fresh-water aquatic environment. This will vary from rapidly moving rivers to static masses of water such as ponds and derelict canals. It will readily be seen that the conditions in slow-running streams will have points in common with those in canals, and there is often a common group of species.

The conditions which influence plant growth in these bodies of water will on the whole be the same as those which operate in the soil. Acidity or alkalinity must play a great part, but now of course there is the more pressing question of the availability of gases—carbon dioxide and oxygen. Access of oxygen to submerged regions is much less free than in terrestrial plants, and this is more the case in still waters than in fast-moving streams where the turbulence will lead to a higher oxygen content. On the other hand, the possibility of damage to the plant structure is greater in fast-flowing water.

As a result it is found that aquatic plants exhibit a number of structural features which are characteristic. The submerged or floating regions have large air spaces for the rapid circulation of gases, whilst many of them show a marked reduction in vascular tissue. Stomata are absent from submerged leaves and appear on the *upper* surface of floating leaves. Submerged leaves are usually small or finely divided—a condition which gives increased surface for absorption and also provides less resistance to water movement. Some of the species have water-borne fruits and seeds, and in some cases the flowers actually open under water, so that the question of pollination becomes a peculiar one.

In a pond or lake the plants show a definite zonation. In the deep regions there may be a few free-floating species such as Duckweeds and Frogbit, and if the water is not too deep, so that there can be adequate light penetration, there will be submerged plants such as the Starworts, Hornwort and some Potamogetons (Pondweeds). Nearer the edge there are species with floating leaves and rhizomes in the mud below. Some of these species will have submerged leaves which differ very markedly from the floating leaves. Amongst other species in this zone will be found the Water-lilies, Water Crowfoot, Floating Pondweed and grasses such as *Glyceria fluitans*. In the shallower margins there will be upright species such as Reedmace (commonly called Bullrush), Arrowhead, Water Plantain, Yellow Irish, Flowering Rush, Purple Loosestrife, Marestail, Burreed, various Rushes and grasses such as Common Reed and *Glyceria aquatica*. Many of these extend shorewards and may occur out of the water, where they are joined by such plants as Balsam, *Mimulus*, Brooklime, Meadowsweet and Marsh Marigold. These latter may be regarded more as marsh plants.

The same zonation may be found along the edge of slow-moving rivers and canals, though the floating species may be absent. In faster waters, particularly the shallower rocky regions, there may be only Algae, Mosses and perhaps Water Crowfoot.

Depending on the nature of the pond margin or river bank, the transition to typically field species may be gradual or abrupt.

Wide swift rivers and all but the margins of deep lakes are without plant life except for small Algae, etc.

Finally we may consider the condition in marshes and bogs. The margins of ponds and rivers often provide a typical marsh, and these may be merged with fenland. Generally they are areas which tend to be basic, and the flora is fairly extensive. Often they are cut by ditches which have a typically aquatic flora.

Besides the many herbaceous species already mentioned, a marsh may have such woody plants as Willows and Alder, and in many cases these trees or shrubs will often be present along the margins. In the ditches or pools can be found the aquatic forms already mentioned, and there may also be Bladderwort floating freely in the water. Many of the species have large rhizomes which are in waterlogged mud, and these have large air spaces in the tissues and adaptations for aeration of the organs. Such plants can tolerate submersion if the water-level rises. At the edge of the marsh the more typical marsh-plants are mixed with those suited to dry land.

There is a decreasing amount of marsh and true fen in Britain as a result of drainage for agricultural purposes, and it may well be that the only natural marshland will be that in National Parks, etc. This type of community is of special interest because it often shelters a wealth of animal life of all kinds, and from the naturalist's viewpoint its loss is much to be regretted.

A bog is typically an acid habitat, so that bogs are generally associated with peaty conditions and are often found on moorlands. They have probably developed on the site of old woodlands in regions of high rainfall, but although water is plentiful, many of the plants show modification to prevent water loss, and this is probably associated with difficulties in absorption and also with a high rate of water loss due to wind, etc. In general, it may be said that bogs are the very wet regions of moorland.

One of the commonest plants is the moss *Sphagnum*, and it may be regarded as the pointer to bog conditions. It may occur in vast masses and is often a principal constituent of peat. In very wet places the *Sphagnum* may actually be in pools, but compared with the standing water of marshes these are usually very poor in both plant and animal life. In these very wet parts one may find along with the *Sphagnum* such plants as Cotton Sedge (often over wide areas), Rushes, Bog Bean, and where the acidity is less there may be Lesser Spearwort and Ragged Robin. Along with these, often on exposed tufts of peat, may be found Cranberry, and on the flatter portions Bog Asphodel, Bog Pimpernel and very typically Sundew, the insectivorous plant. (It is interesting to note that in the low-lying bogs of Ireland the Pitcher Plant, *Sarracenia*, has become established.) There will also be found Sweet Gale, Marsh Pennywort, various sedges and occasionally *Andromeda*. In the less wet patches one gets to the typical moorland association of Heather, Crowberry and Mosses such as *Polytrichum*. The range of species is less than in the marsh, and this restriction may be attributed to the acid conditions which incidentally lead to the development of great depths of peat because there is little or no microbial breakdown of the dead vegetation (coupled possibly with a certain preservative action of the soil constituents). In such circumstances, though possibly in warmer conditions, the accumulation of Carboniferous era vegetation led to the formation of coal.

It must be emphasised that this is only a very brief review of the question of plant associations. In many cases there is a transition from one habitat to another with an overlap of species, and the reader must not expect to go around the countryside seeing sharply demarcated environments, except perhaps actual sheets of water. Moreover, on a continental scale all the different habitats which have been discussed lie within one bigger ecological unit, known as the deciduous forest area. Consideration of such points, however, must be left to specialised works. For the moment it is sufficient to say that by examining half a dozen species in an environment it is often possible to predict what others should be present and also the general character of the soil.

This idea can be expanded by saying that in a particular habitat a few species will always be represented and are not likely to be found elsewhere. A larger number will be present more often in that habitat than in any other, whilst a considerable number of plants will turn up in several environments.

Another point to be borne in mind is that conditions are not necessarily static and that changes are constantly occurring in plant distribution. In a few generations some plants have established themselves in many parts of the country, and we find that Canadian Pondweed (*Elodea*), Balsam and Willow herb are common plants now, but are comparative newcomers. Moreover, the vegetation may be affected by other changes, and in the last two or three years the disappearance of the rabbit from large areas of the country has had a marked effect on many herbaceous plants, partly because grasses and other species are no longer eaten to the point of destruction and partly because the ground is not soured by the droppings.

From another angle it may be said that one of the greatest causes of plant preservation is the retention of the hedgerow as a boundary, involving as it does a margin which is not usually cultivated. Other instances could be quoted, and the student can learn much about plant ecology by watching the changes and successions in the various localities within his range. One thing is certain—a working knowledge of the ecology of a region can be obtained only by constant observation and recording.

One of the greatest ecological problems has been brought about by the development of vast monocultures by man. The increasing demand for food has led to the cultivation of large areas of single crops—even single varieties. This has led to the destruction of the balanced ecology of the natural systems and to the introduction of such problems as epidemic pest attacks by fungi, insects, etc., the impoverishment of the soil because of the lack of return of natural organic material and the demand for intensive application of artificial fertilisers. In some cases this has led to destructive erosion and the contamination of waters, and invariably it has meant the loss of balanced communities. This is an environment and conservational problem, but it cannot be ignored.

6

PLANTS AND MAN

At the beginning of this book reference was made to the importance of plants in the life of man. In this chapter it is proposed to pursue this aspect a little further, and also to draw attention to some of the effects which man's activities have had on the vegetation of the world.

Perhaps the first aspect which might be considered is that of food. No one can afford to ignore man's dependence on plants for many of his staple foods, not only for his own direct consumption but also as fodder for the cattle which provide further major items of diet.

Wheat, Rice, Maize, Potatoes provide a background of nutrition for the greater part of the world's population, and the problem of their maintenance and cultivation provides constant material for research (and worry) by national and international organisations.

In the case of Wheat especially, though not exceptionally, there is a long story of botanical investigation and progress in the production of new varieties which are heavier yielding, more resistant to disease, more suitable for economic bread production and more adaptable to a particular climate. The story of Wheat will always be a fascinating one. Modern Wheats probably arose from species which grew in Asia Minor—it is doubtful whether any one species could be called the ancestor of our modern grain, and it is doubtful whether any of the ancestors still exist in their original form. Compared with the Wheat of today they were small and of poor quality and yield.

The case of Wheat is not an isolated one. A similar kind of development has been followed with all the other food crops, and the Potatoes which we dig from our gardens today are a long way removed from their Andean ancestors.

With the increase in world population the efficiency of food production becomes of ever greater importance.

It follows that in order to increase the production of food crops the natural vegetation has had to be destroyed in order to bring the land under cultivation. Some of the mistakes which were made in the early days of large-scale conversion to arable land have become painfully apparent in modern times. The removal of the natural cover has in some cases led to soil erosion, with the progressive development of what are virtually desert conditions and the loss of valuable land. Hence there has been a revision of methods, and costly steps have had to be taken to preserve arable land for future crop production. Apart from the advantages of climate, one of the reasons why a high degree of fertility has been maintained in the soil of this country is that erosion has not been able to develop because of the small-field system of cultivation and the maintenance of tree-cover. It is true that over wide areas of Britain this method of cultivation is the only one possible, but it has been a valuable restriction which has helped to maintain the fertility of a countryside, which has probably been the most intensively cultivated in the world for the longest period.

Besides the major food crops, however, we must remember that there is an infinite variety of other plant products which are used as food. Fruit of all kinds, tea, coffee, cocoa, vegetable fats are all produced for man's use, and all present their own problems. In some cases the plants concerned require considerable attention and the collection of the crop is a laborious one, as for example tea. On the other hand, the harvesting of the coconut and the maintenance of the trees are much less difficult. In these cases, as in many others, the plant, when once established, will continue to produce crops for a number of years, whereas the banana produces only one crop and is then cut down. This plant, however, is very quick growing and reproduces vegetatively with great ease.

One of the great difficulties which the cultivators have to face is the appearance of diseases which can quickly destroy the whole crop. In fairly recent years we have seen the threat to cocoa production in West Africa from the swollen-shoot disease and the difficulties in combating it (it is an insect-distributed virus), whilst on the other side of the Atlantic great damage was done to banana crops by the Panama disease. Constant vigilance is necessary to offset the effects of such epidemics and to keep ahead of the pests which cause them.

Apart from actual foods, plants are the source of a wide range of spices, condiments and flavouring materials used in the preparation, preservation and seasoning of foods. Though the absolute quantity of some of these substances may not be large compared with some of the staple foods, they are frequently very expensive, and in time past, when other methods of preservation were not known, the spice trade was one of the stimulants to exploration and development, especially in the Far East.

Other problems arise, particularly that of transport. Many of the products are perishable, and for centuries distribution of the world's plant products was restricted, but with improvements in the methods of transport and preservation it is now possible to ensure a much wider consumption of foods from comparatively restricted areas. From the human point of view it means that the failure of food crops in one area no longer means certain starvation.

From plants man still obtains much of the raw material for the making of clothing, etc. The production of cotton is still spreading in the world, and in spite of artificially produced fibres the demand is great. Other fibres are also used: flax for linen-making, sisal and hemp for rope, and in recent years there has been the attempt to produce seaweed fibres—the alginate group. Except in this latter case the plants used for fibre production are cultivated forms and their economic production means careful tending. But in another direction man is still drawing to a very large extent on a natural source of raw material in the form of trees. Every year a tremendous amount of timber is used in the world—this country consumes more than 500 million cubic feet for constructional and other purposes, whilst in the world generally there is a huge drain on forests for the extraction of wood for pulp-making for the paper industry. By far the greater part (about five-sixths) of this timber is softwood, i.e. coniferous, from the northern regions of Europe and America, whilst a considerable amount of hardwood is of tropical or semi-tropical origin. The extraction of timber presents many problems of transportation. There is still a great deal of valuable timber available in the world, but much of it is in relatively inaccessible regions. The demands and difficulties of World War II led to renewed searches for usable timbers, especially in the tropics, and many new woods found their way to this country. With the cessation of softwood control many of these timbers have dropped out again, not least because of the transport problem.

But one problem remains above all others in this connection, and that is the replacement of the trees which have gone. Not only does the loss mean a potential shortage of timber, but it opens the way to erosion and the wastage of valuable lands which could otherwise produce much-needed food. This is now realised in many regions, and in Britain the Forestry Commission aims at producing up to two-thirds of the country's timber requirements, and this will probably involve a productive area of 500 million acres of woodland. It may be pointed out that the need for softwoods is far greater than for hardwoods, and therefore, at least from a commercial point of view, there is an adequate answer to those who criticise the new planting on the ground that so much of it consists of conifers.

Besides wood pulp and cellulose materials, other products such as turpentine and similar substances are obtained from certain coniferous trees.

Another tree product which is still of very great importance despite synthetic production is rubber. Originally rubber was obtained from forest trees in the Amazon region, the trees being natives of the area. Eventually seedlings were surreptitiously removed from the country and planted in Malaya and later in neighbouring countries, where they flourished and gave rise to the very large industry which still exists in that area. The events of World War II revived the South American production again to some extent, but the greatest rival to the South-east Asian production is the synthetic product.

The rubber itself is a latex which flows in special non-vascular elements of the tree (actually a similar substance is produced in many widely separated species) and is tapped by a series of incisions from which the latex drips. The collection of rubber does not involve destruction of the tree, and under proper control the latter can be productive for many years.

A somewhat restricted but nevertheless important industry is the harvesting of cork. This is the bark of *Quercus suber*, a species of Oak growing along the Mediterranean borders. When the tree has reached a certain age the original bark is stripped off to give a fairly even surface, and after that removal of the cork layer takes place at five- to seven-year intervals. The sheets of bark are boiled and subjected to other treatment before the corks and other items can be cut out. The cork-cutting operation requires some skill to ensure that the tree is not damaged.

From the historical aspect at least the medicinal value of plants must rank with the other uses discussed. From the earliest times man has turned to plants for cures for his various diseases, and a great deal of the earlier knowledge of plant structure and products arose from the search for medicinal substances. In fact it is true to say that the first botany books were the herbals in which the properties and appearance of the medicinal plants were described. In spite of the fact that some of the descriptions (and properties!) were highly imaginative, much valuable information was accumulated.

Many of these substances are still extracted, though it is

true to say that synthetic production has replaced extraction in many cases, whilst the development of new and more effective drugs has led to the abandonment of some of the natural substances. In this class we can include quinine obtained from the bark of the Cinchona tree and pre-eminent at one time for the treatment of malaria, but now superseded, for this purpose at any rate, by various synthetic drugs. Though this tendency is spreading as knowledge of the structure of many of the drugs increases, some still defy synthesis, and the natural product must still be obtained. A glance at the list of substances appearing in the wholesale drug markets will reveal that there is still a very considerable production of natural drugs, etc., in the world today. It might be pointed out here that though commercial exploitation of some of these substances has ceased, the plants which produce them are still to be found, and where the products are poisonous the risk to children, cattle, etc., is still a matter to be considered, and care must be taken to prevent children from eating unknown fruits or, of course, known poisonous ones.

Although there has been a decrease in the use of the natural products of some of these plants, there is an increase in the use of another type of substance which has come into prominence in the last two decades. These are the antibiotics, of which the most famous and still most widely used is penicillin. Most of them are produced by Fungi, and in spite of the great publicity attached to the antibiotics, much still remains to be discovered about the range of substances and many of their properties. They have a destructive effect on many pathogenic bacteria even when applied in very small quantities, and for that reason have become very important in medical treatment. Cultivation of the fungi which produce them is therefore necessary on a very large scale and we have the somewhat anomalous position that what was for generations regarded as a troublesome mould on food, etc., has become a plant of vital importance to man.

As with many crop plants, constant improvement of the strain, of the conditions of cultivation and, if one may use the term, of the methods of harvesting is always being developed, so that a higher yield of better quality can be obtained. Although some of these antibiotics are familiar to millions of people, it is possible that we are still at a relatively early stage in our knowledge of them.

Besides antibiotics, Fungi are used to produce such substances as citric acid, gluconic acid, diastase, etc., mostly by controlled fermentations.

But although the nutritive, medicinal and other aspects of plants are so important, there is little doubt that to most people the first appeal of plants is to the eye. It is probable that more species are grown for their ornamental value in flowers or foliage than for all the other purposes together, and much time and money have been spent in improving flowering species. In gardens and parks everyone can see at least some of the thousands of species which are grown for decorative purposes, and the production of flowers for market, either as cut flowers or as bulbs, seeds, etc., for propagation is itself a very large industry. Many species which are seen in gardens and in horticultural displays are exactly as they grow wild in various parts of the world, and have been reared from seeds or plants procured during plant-hunting expeditions, the accounts of which provide fascinating reading. In other cases hybridising and breeding experiments have produced new forms even more attractive than the originals, and there is no doubt whatever that in spite of the economic importance of food crops, etc., man will always turn to flowers for aesthetic relaxation.

No special attention has been given to the question of genetics (the study of heredity) or evolution in this book, but references are given for further reading.

It may be said, however, that the development of plants for economic use has provided an enormous field for research into the problems of genetics. Plants are very suitable for this purpose because they produce large numbers of offspring and relatively frequent generations.

A great deal has been learned about genetics and nuclear behaviour from the efforts of man to improve the strains of plants for his various purposes, and at the present time the chromosome maps of thousands of species are known and their exact genetic constitution recorded. Investigations into the modern aspects of this problem, such as the effects of atomic radiation, are being carried out in many parts of the world, and will doubtless contribute much to this vital consideration for man's future.

A modern problem which causes much controversy and anxiety is the effect of the extensive use of artificial fertilisers, selective weed killers, insecticides etc., on natural communities and it is evident that very serious thought must be given to these effects.

7

FURTHER READING

IT is impossible to provide a full list of suitable books, and those listed below are just a small selection which should be helpful. The reader will doubtless find by his own efforts many which are of equal value, or, in his eyes, better!

GENERAL BOTANY.

Botany of the Living Plant. F. O. Bower. Macmillan.

Textbook of Botany. J. M. Lowson, revised W. O. Howarth and L. G. G. Warne. Univ. Tutorial Press.

Plant Form and Function. F. E. Fritsch and E. J. Salisbury. G. Bell.

Introduction to Botany. J. H. Priestley and L. I. Scott. Longmans.

Intermediate Botany. L. J. F. Brimble. Macmillan.

The Living Plant. P. M. Ray. Holt, Rinehart and Winston.

Organisation in Plants. W. M. M. Baron. Edward Arnold.

PLANT STRUCTURE.

Introduction to Plant Anatomy. A. J. Eames and L. H. McDaniels. McGraw Hill.

Plant Morphology. A. W. Haupt. McGraw Hill.

A Guide to Subcellular Botany. C. A. Stace. Longmans.

PLANT PHYSIOLOGY.

Plant Physiology. W. O. James. Oxford Univ. Press.

An Introduction to the Principles of Plant Physiology. W. Stiles. Methuen.

Plants at Work. F. C. Steward. Addison Wesley.

LOWER PLANTS.

Cryptogramic Botany.
Vol. I. Algae and Fungi.
Vol. II. Bryophyta and Pteridophyta. G. M. Smith. McGraw Hill.

The Fungi. H. G. I. Gwynne Vaughan and B. Barnes. Camb. Univ. Press.

An Introduction to the Biology of Micro-organisms. Hawker, Linton, Folkes and Carlile. Edward Arnold.

British Mosses and Liverworts. E. V. Watson. Camb. Univ. Press.

FLORAS.

Flora of the British Isles. A. R. Clapham, T. G. Tutin and E. F. Warburg. Camb. Univ. Press.

Pocket Guide to Wild Flowers. D. McLintock and R. S. R. Fitter. Collins.

A Concise British Flora in Colour. F. Keble Martin. Michael Joseph.

PLANT COMMUNITIES.

Introduction to Plant Ecology. A. G. Tansley. Geo. Allen and Unwin.

The British Islands and their Vegetation. 2 vols. A. G. Tansley. Camb. Univ. Press.

PRACTICAL WORK.

Textbook of Practical Botany. R. C. McLean and W. R. Ivimey Cook. Longmans.

GENERAL READING.

Plants and Man. R. W. Schery. Geo. Allen and Unwin.

Hormones and Agriculture. G. S. Avery and E. B. Johnson. McGraw Hill.

Trees in Britain. L. J. F. Brimble. Macmillan.

Mushrooms and Toadstools. J. Ramsbotham. New Naturalist Series. Collins.

Collins Guide to Mushrooms and Toadstools. Morten Lange and F. Bayard Hora. Collins.

British Plant Life. W. B. Turrill. New Naturalist Series. Collins.

Mountains and Moorlands. W. H. Pearsall. New Naturalist Series. Collins.

Wild Flowers. J. Gilmour and Max Walter. New Naturalist Series. Collins.

Trees, Woods and Man. H. L. Edlin. New Naturalist Series. Collins.

Poisonous Plants and Fungi. Pamela North. Blandford Press.

Many useful books are now available in the various paperback editions and the student will be able to get further information by cross-references in the literature cited. Some very useful articles on modern developments are available in the publications in the series " Selected Readings from Scientific American " published by W. H. Freeman and Co.

INDEX

* Indicates an illustration